ECO-TOURISM AND ENVIRONMENTAL MANAGEMENT

ECO-TOURISM
AND
ENVIRONMENTAL MANAGEMENT

By

Dr. Govind Prasad
Senior Reader
Dept. of Geography
Rana Pratap Postgraduate College
Sultanpur (U.P.)
(India)

Dr. Shardendu Kislaya
M.Sc., (Chemistry), Ph.D.
Education Instructor
Army Education Crop.
167, Infantry Brigade
Meera Sahab
Jammu

&

Dr. Kanhaiya Lal Gupta
Lecturer
Dept. of Geography
Intermediate College of Varanasi
Varanasi (U.P.)
(India)

DISCOVERY PUBLISHING HOUSE PVT. LTD.
NEW DELHI-110 002

Edition - 2015

ISBN: 978-81-8356-195-2

Eco-Tourism and Environmental Management

Published by:
DISCOVERY PUBLISHING HOUSE PVT. LTD.
4383/4B, Ansari Road, Darya Ganj
New Delhi-110 002 (India)
Phone: +91-11-23279245, 43596064-65
Fax: +91-11-23253475
E-mail: discoverypublishinghouse@gmail.com
sales@discoverypublishinggroup.com
web: www.discoverypublishinggroup.com

Printed at:
Infinity Imaging Systems
Delhi

Preface

The intention of the present Book entitled "Eco-Tourism and Environmental Management" is to lay stress on the operating geomorphic processes and the assemblage of resultant landforms which have fascinated the Panchmarhi hills as an 'Eden of Bliss' and provided the nice balance of natural beauty for the development of tourism industry. Process and form are the synonym of topography and thus the topo-culture has been defined through quantitative techniques and qualitative approaches. Nature attracts the man on their beauty and if the affection is marked too deep, it may cause the arrival of their lovers many times to quench the thirst. In this respect, it is noteworthy if on earth be an eden of bliss, it is Pachmarhi a hill resort, a trekker's paradise, the queen of Satpura and the most verdant jewel of Madhya Pradesh where nature has found exquisite expression in myriad enchanting ways. The observations have been made on the basis of the topographical sheets (Survey of India) interpretation and the extensive field survey of the area. The evaluation of terrain is made through morphometric tools denoting the facts from topo sheets and the problems and prospects have been raised by the judgement of field knowledge. Effects have been made to discuss and explain the topic through scientific approaches. The book comprises seven chapters including sections subsequently. Each chapter separately stands complete in itself but it is interlinked with the succeeding one, presenting a consummate finish with its conclusion.

Chapter 1 includes the study of physio-cultural eco-system wherein the detailed study of topographical characteristics, Physiographic regions, Geological formations and structural growth, drainage system and drainage orientation, climo-ecology, Pedology, flora and fauna sub-surface ecology and groundwater balance and the existing cultural milieu have been mirrored as the basis of the research through the materials available from the Survey of India, District Gazetteer and after the keen observation of the area. The geographical outlook is well represented by cartographical techniques.

Topo-ecology and major morphometric determinants have been discussed in Chapter 2. In view of drainage basin analysis, twelve drainage basins have been selected from different localities and the linear, areal and relief properties of drainage network have been interpretated separately. A critical review on the major components of drainage hierarchy like drainage density (Dd), stream frequency (Sf) and drainage texture (Dt) has been illustrated along with the discussion of relief properties and relief components like average slope (As) and drainage dissection in the last.

Chapter 3 deals with the nature of Morphogenetic processes. The measurement of the intensity of morphological processes have been determined by Peltier's and Wilson's Models. During the course of field observation, the intensity and magnitude of weathering processes and mass movement, chemical weathering, Biotic weathering and fluvial processes have been examined. The nature and effect of operating geomorphic processes are well illustrated in the manuscript.

Landforms are the gift of structure, process and stage. Chapter 4 highlights the origin and development of morphogenetic landforms of the region. Various landforms, for examples, originated due to constant volume weathering, expansion weathering, differential weathering, stripping, limestone weathering and erosional processes have been discussed systematically.

Chapter 5 incorporates the summary account of tourism development in Pachmarhi hill station. The interesting sites of tourist's attraction, causes of tourism improvement, frequency of tourist's arrival, human attitude and role of terrain in tourism development, and the possibilities of area development have also been ornamented in this chapter.

The impact of natural hazards and human interferences on existing ecosystem is the subject of discussion in Chapter 6. The chapter carries the study of the types, trends, dimensions and impact of natural hazards in the region. Human attitude towards existing eco-system, historical review of the crux of eco-degradation, and the Government and public issues for maintaining the eco-problems of the area are the other heads discussed in this column.

Chapter 7 portrays some dialogues of environmental management. In respect to management plans climo-management, Biome management, watershed management, groundwater management and few other kinds of management schemes have been listed in this chapter.

The thesis includes 30 maps and 28 tables. The cartographic representations, generally help in understanding the morphological and cultural complexities of the region.

The author is highly obliged and deeply indebted to Mr. Virendra Kumar and Mr. Jagdish Vishwakarma, Dobhi Jaunpur for drawing the maps used in the book.

The nice co-operation of Smt. Janki Devi, Gitanjali, Surabhi, Shubham, Baby Arya, Mr. Madan Gopal Gupta and Anupam Pandey is appreciable in the completion of the present text.

In the last, but not the least, we are thankful to Mr. T.R. Wasan, Managing Director of Discovery Publishing House, New Dolhi for publishing tho book in duo time.

Teacher's Day
5th September, 2007
Musafirkhana, Sultanpur

Govind Prasad,
Shardendu Kislaya
Kanhaiya Lal Gupta

Contents

Physio-Cultural Ecosystem

PHYSICAL ECOSYSTEM

Eco-system is an ecologic system composed of an organic community (Biotic complex) of plants and animals (Biome) viewed within its physical environment or habitat (Prasad, 1991). It is essentially a somewhat more technical term for 'a segment of nature' and the result of interaction between biological, geochemical and geophysical systems. It is often used in 'Ecology' for the physical background. The study of an 'e' provides a methodological basis for complex synthesis between organisms and their environment.

Eco-system was the term enunciated by 'Tansley' to express 'interacting system' comprising living things and their non-living environment (Tansley, 1935). Stoddart (1965) has developed it as the fundamental organising concept of geography because "Firstly, it is monistic, it brings together environment, man and the plant and animal worlds within a single framework, within which the interaction between the components can be analysed. Secondly, eco-system are structures in a more or less orderly, rational and comprehensible way. The essential fact here, for geography, is that once structures are recognised, they may be investigated and studied. Thirdly, eco-systems, function, they involve continuous throughput of matter and energy. Fourthly, the eco-system is a type of general system and possesses the attributes of general systems. In general, system terms the eco-system in an open system tending toward a steady state under the laws of open system thermodynamics.

Park (1980) advocated that the eco-system stresses the unity of all parts of the environment with morphological processes, it facilitates measurement and comparison of the components of different types of ecological communities and identification of equilibrium and non-equilibrium states; and it highlights basic ecological principles which should be carefully evaluated when guidelines of ecological management and improvement are completed.

In the present study, eco-system is viewed in physical and cultural manner where the physical eco-system involves the summary account of topographical characteristics, physiographic regions, geological formation, drainage system and drainage orientation, climo-ecology, pedology, flora and fauna, sub-surface ecology and groundwater balance of the region. Cultural eco-system includes an up-to-date knowledge of man-made activities and the resultant culture in brief.

Location, Extent and Limitation

The region (Pachmarhi Hills) popularly known as 'Panchmarhi' is located in between 22° 16' 48" N and 22° 36' N latitudes and 78° 10' E and 78° 38' E longitudes. From east to west, it is extended in 24' longitude (about 41 kms of length) and from north to south it lies in 19' 12" latitude (about 35 kms of width). The maximum area of the region (about 770 sq.kms.) is located in the south-eastern part of Sohagpur Tahsil of Hoshangabad district (M.P.). A limited area marked in the north-east corner of Betul tahsil of Betul district (about 33.50 sq.kms.) and in the north-west portion of Jamai tahsil of Chhindwara district (104.50 sq.kms.) is also included along the southern boundary of the region. The total geographical area of the region is 908 sq.kms. which falls in topographical Sheet Nos. 55J/2, 55J/3, 55J/6, 55J/7, 55J/10 and 55J/11 of Survey of India (Fig.1.1). The region is located in Malwa region in India.

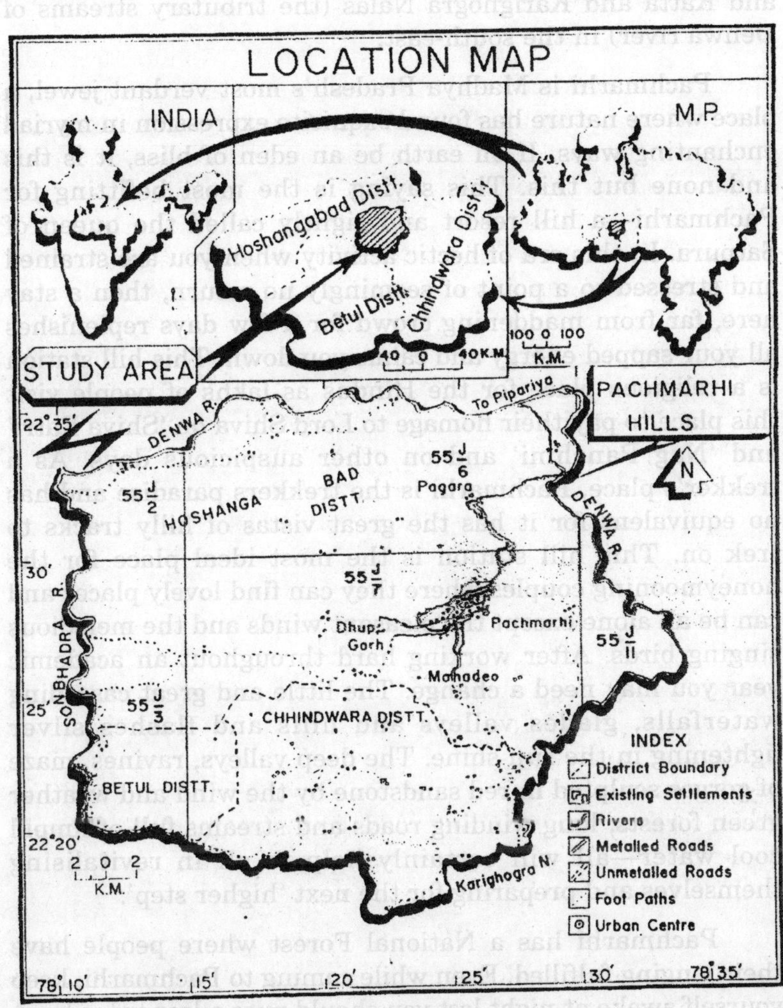
LOCATION MAP
INDIA
M.P.
Hoshangabad Distt.
Chhindwara Distt.
Betul Distt.
STUDY AREA
40 0 40 K.M.
100 0 100
K.M.
PACHMARHI
HILLS
To Pipariya
DENWA R.
DENWA R.
N
22° 35'
55 J/2
HOSHANGA
BAD
DISTT.
55 J/6
Pagara
30'
55 J/7
Dhup Garh
Pachmarhi
55 J/11
Mahadeo
SONBHADRA R.
25'
55 J/3
CHHINDWARA DISTT.
INDEX
District Boundary
Existing Settlements
Rivers
Metalled Roads
Unmetalled Roads
Foot Paths
Urban Centre
BETUL DISTT.
22°20'
2 0 2
K.M.
Karighogra N.
78° 10'
15'
20'
25'
30'
78° 35'

Fig. 1.1

Pachmarhi hills are bounded by the Denwa river in the north and east, Sonbhadra river in the west and south-west and Katta and Karighogra Nalas (the tributary streams of Denwa river) in the south-east.

Pachmarhi is Madhya Pradesh's most verdant jewel, a place where nature has found exquisite expression in myriad enchanting ways. If on earth be an eden of bliss, it is this and none but this. This saying is the most befitting for Pachmarhi—a hill resort and rightly called the queen of Satpura. In this era of hectic activity when you are strained and stressed to a point of seemingly no return, then a stay here, far from maddening crowd for a few days replenishes all your sapped energy and calms you down. This hill station is a religious place for the Hindus as lakhs of people visit this place to pay their homage to Lord Shiva on 'Shiva Ratri' and 'Nag Panchmi' and on other auspicious days. As a trekker's place, Pachmarhi is the trekkers paradise and has no equivalent for it has the great vistas of hilly tracks to trek on. This hill station is the most ideal place for the honeymooning couples where they can find lovely places and can be all alone except the pleasant winds and the melodious singing birds. After working hard throughout an academic year you may need a change. The little and great cascading waterfalls, glades valleys and hills and flashes silver lightening in the sun shine. The deep valleys, ravines, maze of gorges sculpted in red sandstone by the wind and weather green forests, long winding roads and streams full of limpid cool water—all will certainly help to all in revitalising themselves and preparing for the next 'higher step'.

Pachmarhi has a National Forest where people have their longing fulfilled. Even while coming to Pachmarhi, keep yourself awake at night lest you should miss a leopard, a bear, a bison or some other wild animals lurking in the vicinity or walking majestically across the road. And if you just want to do nothing and be in tune with infinity, Pachmarhi would not fail you in this respect too. Just stroll along some distance away from your lodging and you get the environment which had been described and longed for by 'Rishis' of ancient times for prayer and meditation.

Pachmarhi presents a nice picture of topo-features where the setting sun colours the green leaves by its rays, scattered hills standing in silent prayers to the power unseen, waterfalls giving an allusion of bees buzzing, deep ravines and valleys to remind us how deep depth could be, dark silent caves where we could feel the coolness of silence and rediscover ourself. At night one can listen to the sighs of swaying trees.

The original name of Pachmarhi was 'Panchmarhi' based on Panch Pandavas, which can be traced back to the period of Mahabharat. Historical evidences advocate that Panch Pandavas have passed their exile period in the Pandav caves of Pachmarhi. In Mahabharat, it was named as 'Khandav van' the place of security of Pandavas and at that time it was well recognised as 'Agyat Bas' landmass for Pandavas.

Pachmarhi is a large plateau ringed by hills on the ever-green central Satpura ranges. Satpura may be defined as 'Sat Pura', the seven sons of the Vindhyachal mountain or 'Satpuda' (the seven folds). Pachmarhi hills is also known as Mahadeo hills. The name 'Mahadeva' especially denotes a single peak, 1336 metres high at the southern edge of the small but raised tableland of Pachmarhi, at the feet of which are the sacred cave and the shrine of Mahadeo. The Mahadeo hills was originally applied to include the widermass of sand-stone hills which is encircled by the Denwa and the Sonbhadra rivers. Now, it includes the whole section of the Satpura mountain between the Chhindwara plateau and the Narmada valley. It includes several parallel ranges and their branches at flanks from the Tawa in the west to the Dudhi in the east. Thus, the Pachmarhi plateau including hills is a part and parcel of Mahadeo hills of Satpura mountain ranges.

Topographical Characteristics and Physiograph Regions

Pachmarhi hills are an essential part of the Sa range. As discussed earlier Satpura is treated as the sons of the Vindhyachal mountain or Satpuda (the seven folds), referring to the numerous parallel ridges. It is the name applied for a far more lengthy area, about a thousand kilometres, with homogenous nature commencing from Amarkantak in the east to the proximity of the western coast (Sinha, 1994). The whole mass of Satpura hill ranges is divided by distinct names to each separate block which, from height or sacredness, has acquired some degree of notoriety. The major chains of Satpura are named as the Raj Pipla, Kalibhit, Asirgarh, Satpura proper, Mahadeo, Maikal and Saletekri. The ranges between the Mahadeo and the Maikal are generally referred by the names of local features of importance.

Pachmarhi hills depict the assemblage of land-forms. It is a smaller part encircled by Denwa and Sonbhadra rivers from all side. It is an uplifted land mass mostly showing massive sandstone. The raised tableland (Plateau) of Pachmarhi table is ornamented with table upon table topography, Mesas and Buttes, Springs, caves, narrow cut deep gorges, deep drained rivers, typical ravines, rapids and waterfalls, isolated dissected hills from all sides and boulders controlled valleys. The topography of the region is very complex in nature. The plateau is highly dissected by the Denwa river in the north and west which separates the low outer (northern most) range of Satpura. In the west, the upland is deeply notched down by the Sonbhadra river, a tributary stream of the Denwa river. Dhupgarh (1352m) is the highest point between the Nilgiris and the Himalayas except Mount Abu. At places the superficial stratum on the top of the hills is trappean but a Pachmarhi hills, sandstones are uncovered; though the cristalline rocks are the upper most cover on the Satpura elsewhere. The northern face of the plateau is distinct from the plain by the sudden uplift.

The highest elevation of the region is marked by 1350m of contour which shows the highest peak of Dhupgarh Pahar. The lowest elevation is noticed in the north-west flank along the Denwa and Sonbhadra rivers (Near the confluence point of Denwa and Sonbhadra rivers) by 340m contour. Almost the border line of the region is demarcated by rivers so that the area along the drainage lines (borderline) have shown the lowest elevation of the region. The height of the plateau increases towards the heartland of the Pachmarhi. In the south and southeastern portion, the spot height of 580m is registered along the Karighogra nala while in the south-west border line along the valley region of Sonbhadra river, local height is illustrated as 500m. In the extreme north-east and west, along the river Denwa, lowest elevation is found as 420m, 500m and 520m registered through corresponding contour lines (Fig.1.2).

Fig.1.2 depicts the main contour lines of the region showing topographical characteristics wherein prominent hills of the region have been also mirrored. The major hills of the region are like Dhupgarh Pahar (1352m), Mahadeo Pahar (1330m), Chauragarh Pahar (1308m), Titanga Pahar (1240m), Pandidev Pahar (1160m), Richkhoh Pahar (1160m), Belkhandar Pahar (1150m), Jatashanker Pahar (1139m), Barkachhar Pahar (1108m) Burimai Pahar (1092m), Jambudip Pahar (1090m), Handikhoh Pahar (1084m), Kudardev Pahar (1083m), Chinchaman Pahar (1070m), Nandigarh Pahar (1060m), Marodev Pahar (1060m), Panch Pandav (1060m), Khamkhera Pahar (1040m), Brijlaldeo Pahar (1008m), Dadar Pahar (975m), Bharna Pahar (960m), Nimdhana Pahar (947m), Somgarh Pahar (900m), Nishangarh Pahar (900m), Ghughudev Pahar. (900m), Patarkota Pahar (893m), Tatragarh Pahar (880m), Mahuldev Pahar (873m), Gauriyadev Pahar (860m), Barghat Pahar (834m) Lambitangi Pahar (762m), Mankidev Pahar (760m), Malulmatta Pahar (732m), Durandiya Pahar (720m), Kiwarkhap Pahar (700m), Anjan Toriya Pahar (613m), Chuchara Pahar (590m), and Langi Pahar (580m) respectively.

Fig.1.3 shows the cross section of major hilly country of the region and denotes the relief irregularities everywhere. The cross section of Patarkota Pahar (Fig.1.3.1), Mankidev,

PACHMARHI HILLS
CONTOURS MAP
DENWA R.
SONBHADRA N.
KARIGHOGRA N.
5 0 5
K.M.
LIST OF PAHAR (Hills)
1–Somgarh 935M
2–Dadar 975M
3–Gauriyadeo 860M
4–Malulmatta 732M
5–Chuchara 590M
6–Langi 580M
7–Anjantoriya 613M
8–Nimdham 947M
9–Patarkota 893M
10–Belkhandar 1150M
11–Tatragarh 880M
12–Nanoligarh 1060M
13–Barkachhar 1108M
14–Jambupip 1090M
15–Marodeo 1060M
16–Khakhera 1040M
17–Jatashankar 1134M
18–Pandidev 1160M
19–Panchpandav 1060M
20–Bharna 960M
21–Dhupgarh 1352M
22–Niehangarhta 900M
23–Chintaman 1070M
24–Brijlaldev 1008M
25–Mankidev 760M
26–Lambitangi 760M
27–Barghat 834M
28–Richhkhoh 1160M
29–Titanga 1240M
30–Mahadev 1330M
31–Chauragarh 1308M
32–Handikhoh 1084M
33–Duranandiya 720M
34–Kiwarkhap 700M
35–Kudardeo 1083M
36–Mhauldev 873M
37–Ghughudev 900M
38–Burimai 1092M

Fig. 1.2

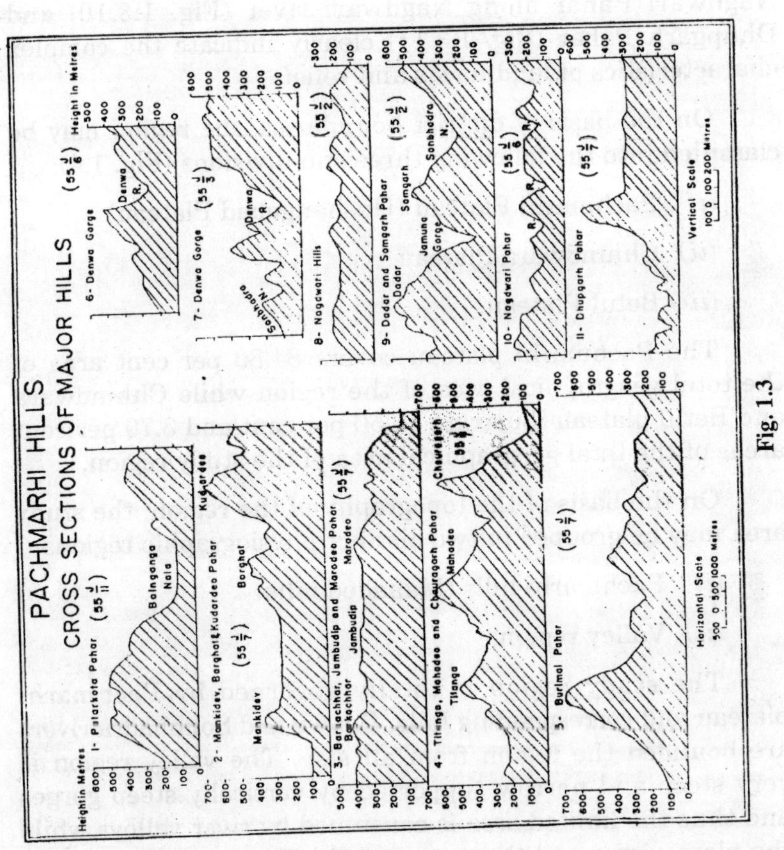
PACHMARHI HILLS
CROSS SECTIONS OF MAJOR HILLS
1- Palarkota Pahar
Bainganga Nala
2- Mankidev, Barghat & Kudardeo Pahar
Mankidev
Barghat
Kudardeo
3- Barsachhar, Jambudip and Marodeo Pahar
Barsachhar
Jambudip
Marodeo
4- Tilanga, Mahadeo and Chauragarh Pahar
Tilanga
Mahadeo
Chauragarh
5- Burimal Pahar
6- Denwa Gorge
Denwa R.
7- Denwa Gorge
Sonbhadra N.
Denwa R.
8- Nagdwari Hills
9- Dadar and Samgarh Pahar
Dadar
Jamuna Gorge
Samgarh
Sonbhadra N.
10- Nagdwari Pahar
11- Dhupgarh Pahar
Height in Metres
Horizontal Scale
Vertical Scale

Fig. 1.3

Barghat and Kudardeo Pahar (Fig.1.3.2), Barkachhar, Jambudip and Marodeo Pahar (Fig.1.3.3), Titanga, Mahadeo and Chauragarh Pahar (Fig.1.3.4), Burimai Pahar (Fig.1.3.5), hill cross section along Denwa river (Fig.1.3.6), along Sonbhadra river (Fig.1.3.7) along Sangwa Nala (Fig.1.3.8), Dadar and Somgarh Pahar along Sonbhadra river (Fig.1.3.9), Nagdwari Pahar along Nagdwari river (Fig. 1.3.10) and Dhupgarh Pahar (Fig. 1.3.11) clearly indicate the complex characteristics of land-forms and relief.

On the basis of district boundaries, the region may be classified into the following three sub-divisions (Fig. 1.4):

(i) Pachmarhi Plateau (Hoshangabad Plateau)

(ii) Chhindwara Plateau

(iii) Betul Plateau.

The Pachmarhi plateau covers 84.80 per cent area of the total geographical area of the region while Chhindwara and Betul plateau illustrate 11.50 per cent and 3.70 per cent areas of the total geographical area of the study region.

On the basis of the topography of the region, the study area may be grouped in two distinct physiographic regions.

(i) Pachmarhi hills (Mahadeo hills)

(ii) Valley region.

The study region is totally governed by Pachmarhi plateau and corresponding hills. Denwa and Sonbhadra rivers are bounded the region from all side. The valley region is very steep and narrow supported by vertically steep gorges and thus the limited area is consumed by river valleys while the plateau rim and the surrounded hills occupy the greater portion of the region.

Geological Formation and Structural Growth

Following the geological notes included in District Gazetteers of Hoshangabad, Chhindwara (Sinha, 1994, 1995) and Betul districts (Srivastava, 1990), the geology of the

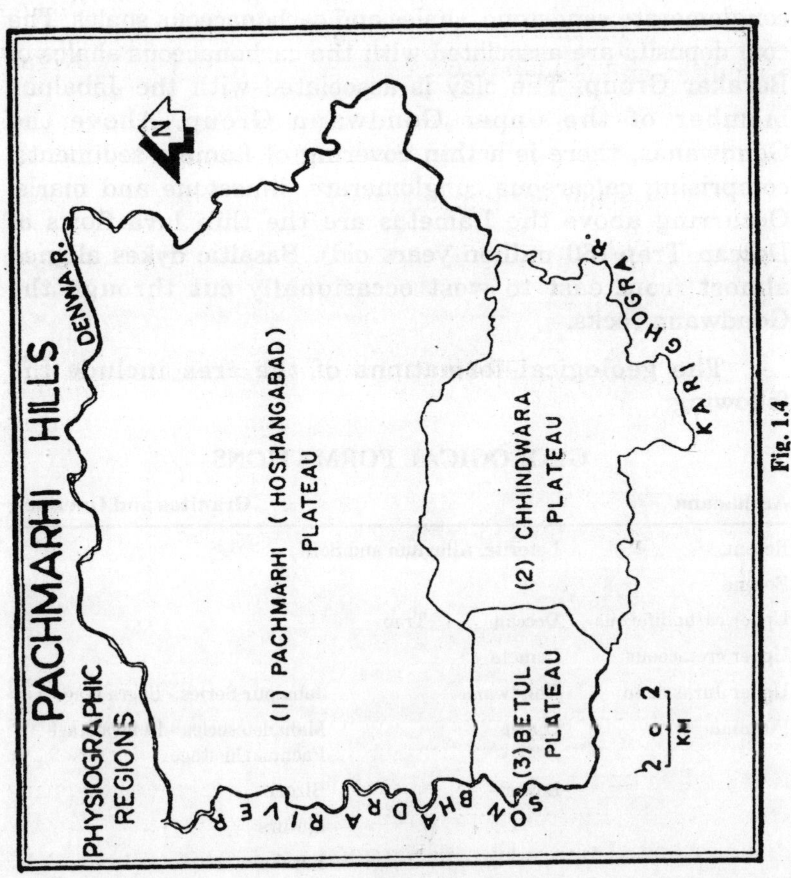
PACHMARHI HILLS
PHYSIOGRAPHIC REGIONS
N
DENWA R.
(1) PACHMARHI (HOSHANGABAD)
PLATEAU
(2) CHHINDWARA
PLATEAU
(3) BETUL
PLATEAU
SON BHADRA RIVER
KARI GHOGRA R.
2 0 2
KM

Fig. 1.4

region is discussed. The region generally occupies the Gondwana rocks. Other rocks are rarely seen. The region is occupied by the Archaean rocks (2500 million years old) comprising granitic gneiss, green schists, intrusive granites, basic rocks, quartz veins etc. Above the archaeans are the Gondwana sediments (200 million years old) which include conglomerate sandstone, shales and carbonaceous shales. The coal deposits are associated with the carbonaceous shales of Barakar Group. The clay is associated with the Jabalpur member of the upper Gondwana Group. Above the Gondwanas, there is a thin covering of Lameta sediments, comprising calcareous conglomerate, limestone and marls. Occurring above the Lametas are the thin lava flows of Deccan Trap (60 million years old). Basaltic dykes aligned almost from east to west occasionally cut through the Gondwana rocks.

The geological formations of the area include the following:

GEOLOGICAL FORMATIONS

Archaeans			**Granites and Gneisses**
Recent	Laterite, Alluvium and Soil		
Eocene			
Upper carboniferous	Deccan	Trap	
Upper cretaceous	Lameta		
Upper Jurassic to	Gondwana		Jabalpur Series - Bagra stage
Permian	Upper		Mahadeo series - Denwa stage Pachmarhi stage
	Lower -		Bizori
			Moturs
			Barakars
			Talchirs
Bijawars			Quartzites, cherts breccia, dolomite etc.
....	Unconfirmity		....
Archaeans	Gneisses, etc.		Granites, Phyllites, Schists

Archaeans

The Archaeans rocks, the oldest formation are present as interrupted bands along the southern edge of Narmada—alluvium from the eastern end to westwards of Hoshangabad district. The maximum width of the interruped bands is about 10 km but generally it is less. The Archaean rocks generally consist of gneissic granites, phylites, Schists, quartzites and a calcareous group of rocks. The phylites and Schists are always closely related to the phylites and Schists with which they grade imperceptably. The Archaeans are also present predominantly in the west of alluvial tract between Harda and western boundary of Hoshangabad district which is out side of the study region. This is represented mainly by gneissic granites, is massive porphyritic, grey to pink in colour and contains bands and lenses of amphibolite. Lenticular quartzite bands are occassionally present. These rocks are unconformably overlain by Bijawar groups of rocks. The area occasionally experiences the above formation.

Bijawar Group

The Bijawars occur as a thick band of about 18 km going westwards from Harda. This band bifurcates into two individual bands at about 2 km west of Harda-Handia road and continues westward as individual bodies. The Bijawar rocks consist of quartzites which form isolated hills parallel to Narmada river, quartzitic and chert breccia which forms extensive unit and dolomite which occurs in an area of 15 km. The rocks are folded with east-north-east-west-southwest axis.

Gondwanas

Pachmarhi hills generally show the Gondawana group of rocks. In the south of the Narmada valley the Gondwana rocks are fully represented. These may be about 2400m thick at the south of Pachmarhi plateau. Stratigraphically the lower most rocks are represented by Talchir shales and sandstones overlain by sandstone of Barakar stage. The Barakars are overlain by a succession of rocks which have been divided by two stages, the Motur below the Hijori above. The Moturs consist of thick soft or coarse beds of grey and brown sandstone with occasional lenses of sandy clay and shale which are very rarely calcareous. The sandstone are massive and contains grits. The clay of Moturs stage is generally calcareous. Scarce plant fossils are presented in the shales and carbonaceous or silicified tree trunks. The rocks of Bijori stage generally occur at the south base and extend westwards upto some distance west of Hoshangabad-Itarsi road. The rocks of this stage generally comprise shale which are occasionally carbonaceous and micaceous flags and sandstone. Minor coal seam are sometime present. Plant fossils and occasional vertebrate animal fossils have been recorded from these beds.

Overlying rocks of Mahadeo series belonging to the upper Gondwanas are represented by thick Pachmarhi sandstone sequence with an unconformity in between. The Mahadeo series is illustrated by Pachmarhi, Denwa and Bagra stages. The typical Pachmarhi beds are uniform in character and consist of coarse, white and usually soft sandstones. As a rule stratification is observed by oblique lamination and both partings and bedding planes are generally characterized by iron infiltration. The Pachmarhi sandstones occupy high ranges of hills and give rise to vertical scarps of great height. The Pachmarhi beds dip northwards at a lower angle and are overlain by Denwa beds north of Pachmarhi Plateau. These beds are comprised principally of soft variegated clays, thick beds of which are interbedded with lenticular white sandstone beds corresponding grit. The Denwa clays are always calcareous and the colour varies between grey, red and buff. The Denwa beds have yielded vertebrate fossils. This stage is overlain by Bagra stage and occurs at the southern margin of the Gondwana terrain. In the western part overlap the Denwa beds and Pachmarhi sandstones and eastwards they pass laterally into conglomerates which becomes the principal constituents.

Overlying the rocks of Mahadeo series is the Jabalpur series which in Pachmarhi Plateau of massive sandstone alternates with soft white clay subordinate bands of conglomerate are the haematitic coal carbonaceous shale. Red clay and chert may also occur. The sandstones are very soft. The Jabalpur cover major parts of Budhimai reserved forest southeast of Seoni-Malwa and Bagra reserve forest.

Lameta

Near Kaveli, Amadad and Kirathpur villages, occur some cherty limestone with high percentage of calcium. These are overlain by the Deccan Trap formation.

Deccan Traps

Basaltic lava flows of Deccan Trap are mostly seen in the southwestern as well as northeastern corner of Hoshangabad district. Small patches of Deccan Trap are seen overlying in the Satpura range. The Trap rocks comprise of almost horizontal lava flows which on weathering have given rise to black cotton soil. Basaltic dykes with a general east-northeast west-southwest trend and connected with the Deccan Trap erruption are found to intersect the earlier rocks as well as some flows of Deccan Trap especially in southern part of the area.

Laterite

Laterite forms a cover on the Trap formations, generally on hills and mounds. These have resulted from the effects of weather and water.

Narmada Alluvium

More than half of the area of Hoshangabad district is occupied by the Narmada alluvium which consists stiff reddish to yellowish or brownish clay with numerous interrelations of sand and gravel. The particles of Narmada alluvium are also seen in the study region.

Thussu and Gyan Prakash (1987) have also interpretated the formation of rocks of Gondwana Super group in Pachmarhi and its surrounding areas (Fig.1.5A). Spate (1954) feels that the Deccan lavas are practically horizontal and remarkably homogeneous basalts were probably extruded from fissures towards the end of cretaceous, though a flora which seems to be of early Eocene age is found between some of the flows. The lavas were poured out on to a land surface which had already attained an advanced stage of maturity and form a most striking feature in the geomorphology of the Peninsula. Ayyar, Verma, Nigam and Singh (1971) have stated that the upper Gondawana sandstones of the Mahadeo series on weathering take peculiar forms resembling bastions, buttresses and battlements and are cut by deep canyons.

The Pachmarhi hills was the area previously covered by Satpura basin (Fig.1.5B). The region was well affected by Narmada rift zone in the north and Asirgarh gap in the south. The Palaeographic map (Fig.1.5B) shows dominant drainage from Permian to Triassic in the Son valley and Satpura basin. The drainage pattern suggests the possible continuity of the Son-Mahanadi basin with the Satpura basin and its extension further west underneath the Deccan Traps to the south of the Narmada river (Casshyap, 1982). It is suggested, on the basis of palaeodrainage pattern that the Gondwana rocks of the Satpura basin must have extended westward beneath the traps with a possible outlet of the Gondwana rivers also in that direction. Casshyap also advocated that the Mesozoic palaeocurrent patterns in the Pachmarhi sandstone in the central part of the Satpura basin are west-southwest and west-northwest directed in close correspondence with the Permian Palaeocurrents of the western parts of the Son valley basin. The Permian Gondwana drainage in the Satpura basin likewise records a dominance of westerly flow component. The Palaeocurrent pattern, in deep favours westward continuity of the Satpura basin to the south of Narmada valley (Narmada lineament) underneath the thick pile of Traps. Consequently, the lack of Gondwana outcrops hampers direct determination of a westerly outlet of Gondwana drainage in this critical part of western Peninsular India. Fig.1.5B supports the existence of Precambrian highlands not far to the north of Son valley and

Satpura-Gondwana basins. However these highlands contributed sediments locally and occasionally both to the Permian and Mesozoic Gondwana basins (Casshyap, 1977).

Valdiya (1982) stated that the Satpura basin feels tensional tectonics. A tectonic event of epeirogenic nature not only terminated the process of sedimentary accumulation but gave rise to long, deep normal faults represented by the Son-Narmada faults, the Great Boundary Fault of Rajasthan and similar minor faults as shown in Fig.1.5A. Under the northerly sloping cration, hot spots developed towards the close of the Palaeozoic giving rise to a triple junction, characterized by all aborted rifts, down which flowed rivers in the northwesterly directions and deposited thick continental Gondwana sediments. At the end of the Mesozoic and the beginning of Cenozoic, there was a great outpouring of lavas, giving rise to the Deccan Trap.

Qureshy (1964) discussed that a broad gravity—high characterizes the Satpura range (Fig.1.5C). The relief 750 mgl brings out clearly the existence of horst and graben topography in the western side—the Satpura forming the horst and the Narmada river occupying the graben.

Drainage System and Drainage Orientation

Pachmarhi hills represent an unique type of drainage system which constitutes the most important element of surface geodynamics. In general, rivers and valleys have a special places for it is impossible to treat the development of landforms, or to describe existing forms in a rational manner without constant reference to the valleys that have been worn in them and to the rivers by which the waste is washed along the channel in the valley floor (Davis, 1900). Drainage system is highly controlled by geological structure, local relief, slopes and other physical attributes which bear responsibility to form in a new shape to the country concerned. In broad sense, the drainage of the region, so epitomizes the history of regions, physiography as the history of its rivers (Fennemen, 1938).

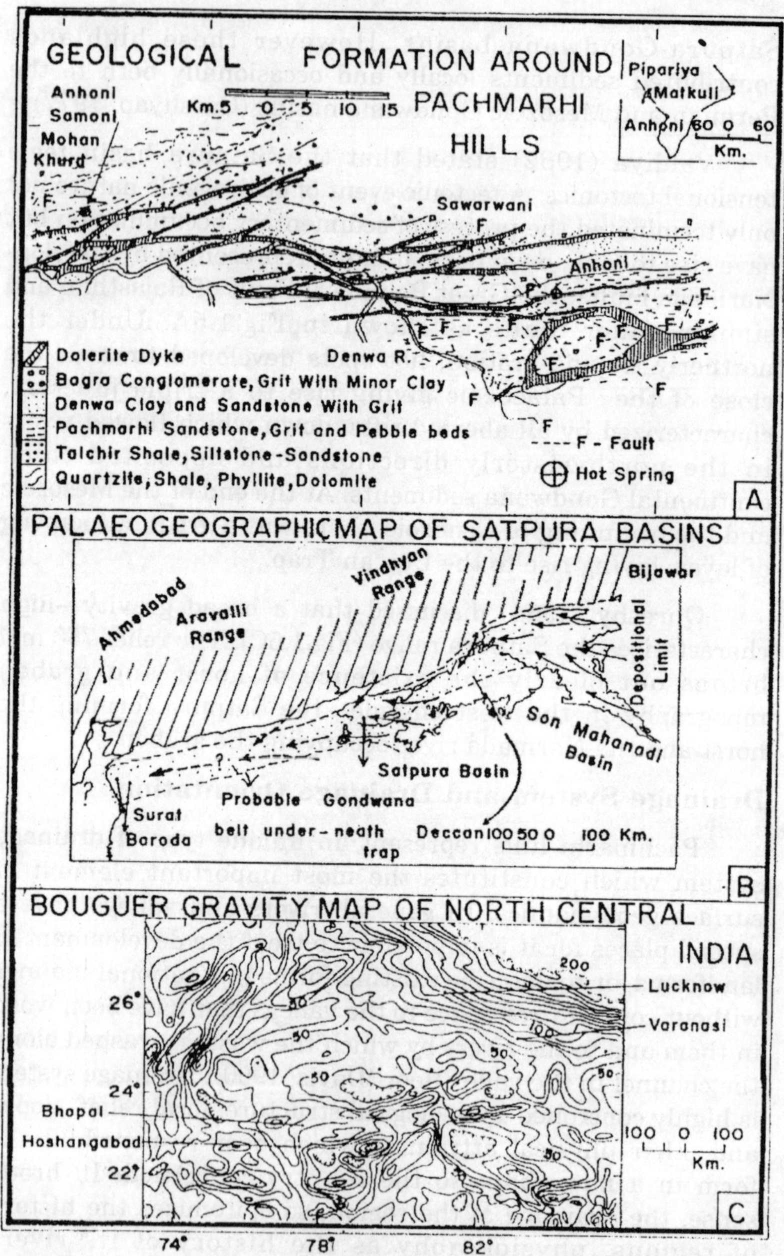
GEOLOGICAL FORMATION AROUND PACHMARHI HILLS
Km.5 0 5 10 15
Anhoni
Samoni
Mohan
Khurd
Pipariya
Matkuli
Anhoni 60 0 60
Km.
Sawarwani
Anhoni
Denwa R.
Dolerite Dyke
Bagra Conglomerate, Grit With Minor Clay
Denwa Clay and Sandstone With Grit
Pachmarhi Sandstone, Grit and Pebble beds
Talchir Shale, Siltstone-Sandstone
Quartzite, Shale, Phyllite, Dolomite
F - Fault
Hot Spring
A
PALAEOGEOGRAPHIC MAP OF SATPURA BASINS
Ahmedabad
Aravalli Range
Vindhyan Range
Bijawar
Depositional Limit
Son Mahanadi Basin
Satpura Basin
Probable Gondwana belt under-neath trap
Surat
Baroda
Deccan 100 50 0 100 Km.
B
BOUGUER GRAVITY MAP OF NORTH CENTRAL INDIA
200
Lucknow
Varanasi
100
50
26°
24°
22°
Bhopal
Hoshangabad
100 0 100
Km.
74°
78°
82°
C

Fig. 1.5

Pachmarhi hills are drained by the river Denwa and its major tributary stream Sonbhadra from all sides. Denwa forms the major drainage system of the region. Sonbhadra drainage system is the second major drainage system of the region. Among the complex of hills around Pachmarhi Plateau a line of hills extends from Dhupgarh (1352m) to Burimai Pahar (1092m) located about 15 km in the south. Their eastern flow into the Denwa and the western into the Sonbhadra. The Denwa rises at Denwa Khud south of the plateau and circuits it flowing to the south, east, north and the west. The circle is complete by the course of the Sonbhadra which joins the Denwa after flowing first to the west and then to the north. The Bori is a small tributary of the Sonbhadra flowing west. The name is famous for the excellent teak forests in its valley. The Denwa carves a narrow valley extending east to west between the two northern ranges of Mahadeo hills. The Denwa valley is separated from the Narmada Valley proper. The Denwa joins the Tawa near Bagra and is an equally important stream. The Tawa bridge on the Piparia-Pachmarhi road is frequently covered by floods during the rains for a few hours but seldom for longer. Nagdwari nadi, Bainganga nala, Mechidhar nala, Sawania nala and Ganjakunwar nala are the other minor tributary streams of the Denwa river. Bori, Jhuli, and Jamna nala are the subtributaries of Sonbhadra river. The Bori drainage system includes the Nishan and Kabra tributaries in its catchment area. Mostly all the drainage lines change their direction according to lithological controls of the region. The rivers also make their sinuous course due to structural control and the relief existed there. The length of the drainage lines draining over the Pachmarhi hills like Denwa, Sonbhadra, Nadwari, Kabra, Nishan, Bainganga, Bori, Ganja Kunwar, Machidhar, Sawania, Jamna, Kuma-Jhiri and Jhuli have been measured as 109.50 km, 59.00 km, 33.00 km, 24.10 km, 18.80 km, 15.20 km, 14.60 km, 14.00 km, 13.10 km, 13.00 km, 8.70 km, 8.50 km and 7.50 km respectively.

The drainage system of the region is shown in Fig.1.6 where the catchment area of the drainage basins indicates

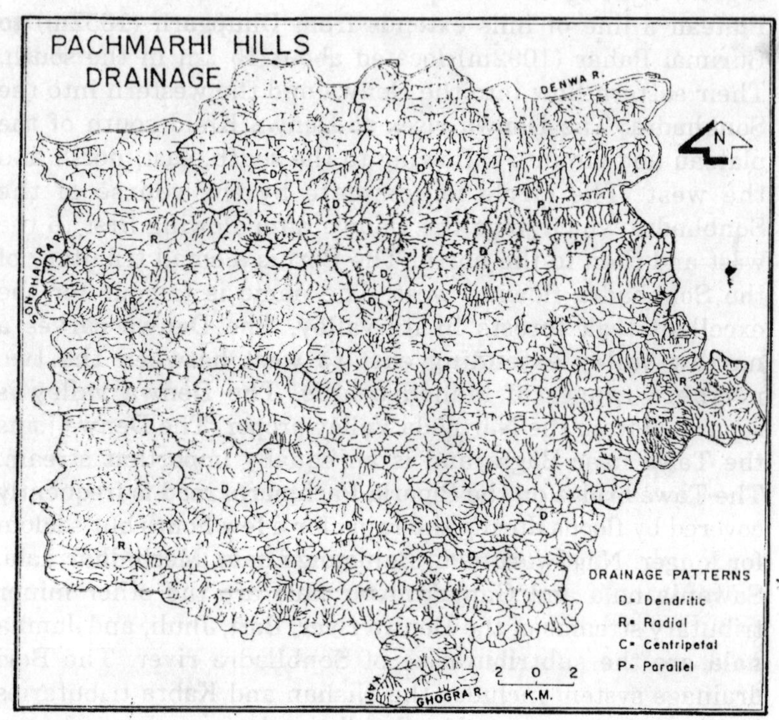
PACHMARHI HILLS
DRAINAGE
DENWA R.
N
SONBHADRA R.
DRAINAGE PATTERNS
D- Dendritic
R- Radial
C- Centripetal
P- Parallel
2 0 2
K.M.
GHOGRA R.

Fig. 1.6

the water dividing lines in the region. The map also shows the major drainage patterns of the region along with springs, waterfalls, rapids, and caves surveyed in the field and recognised on the topographical sheets. The most encountered drainage pattern is the dendritic which is characterized by irregular branching of the tributary streams in many directions and almost at any angle although usually at considerably less than a right angle (Prasad, 1986). Parallel, Radial, Rectangular and Centripetal are the other drainage patterns observed in the region. The study region indicates 4296 streams of Ist order, 946 streams of IInd order, 185 streams of IIIrd order, 44 streams of IVth order, 9 streams of Vth order and only one stream of VIth order respectively. Thus, the Denwa river is a VIth order stream.

The drainage orientation (Fig. 1.7) is influenced by the lithological control, slopes and topographic variations of the region. Most of the river along with tributaries are formed due to tectonic movement and their paths are much guided by linear deep gorges. On the basis of the Figs.1.6 and 1.7, it may be said that the direction of the river is east to west in the east and west to east in the west. The Denwa and Sonbhadra rivers show the flow direction generally from south to north, but in the north, the Denwa river flows from east to west. All these facts validate the fact that there are much dominance of lithology and the region is experienced the tensional tectonics.

Climo-ecology

The region experiences a dry climate except during the southwest monsoon season. The year may be divided into four periods:

(i) The cold weather season—December to February

(ii) The hot season—March to second week of June

(iii) The monsoon season—Mid June to end of September

(iv) The post-monsoon season—October and November.

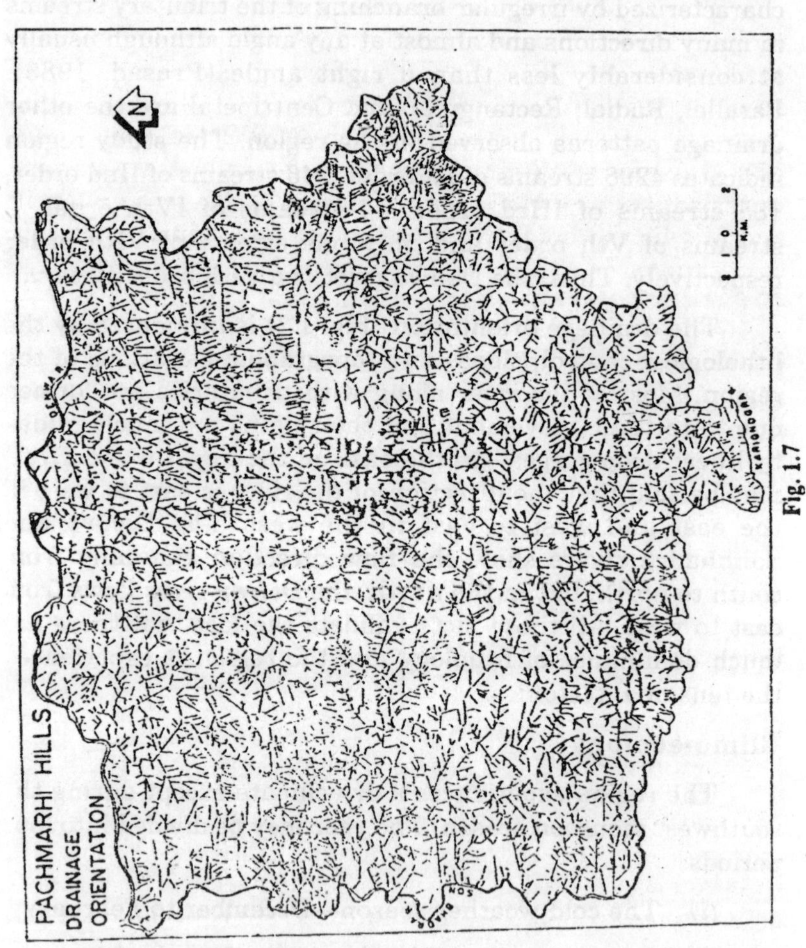

Fig. 1.7

The climatic conditions of the region is shown in Fig.1.8, and Tables 1.1, 1.2 and 1.3. The detailed discussion of climo-ecology may be presented in the following manner.

Temperature

The average annual maximum and minimum temperature of Pachmarhi (Fig.1.8A and Table 1.1) are observed as 26.6°C and 16.6°C respectively. The normal annual temperature is recorded as 21.76°C. The maximum daily temperature ranges between 21.6°C (January) and 35.6°C (May) while minimum average daily temperature varies between 7.7°C (December) and 24°C (May). Thus April and May is the hottest and December and January are the coldest months of the region. The highest temperature as 40.6°C (recorded on 30 April 1942 and 4 to 8 May 1954) is recorded in the month of April and May. The lowest temperature as 0.6°C is recorded on 2 February 1928. The average highest and lowest temperature of the region have been registered as 34.10°C and 8.2°C respectively. The summer season becomes pleasant although the daily temperature in the summer season may sometimes go up above 40°C. Hot dust-laden winds which blow during the summer season add to the discomfort in the plains of Hoshangabad. Afternoon thundershowers which occur on some days bring welcome relief though only temporarily. With the onset of the southwest monsoon by about mid-June, there is an appreciable drop in temperatures. With the withdrawal of the monsoon after about the end of September there is a slight increase in day temperature in October but the nights become progressively cooler. From about mid-November both day and night temperature decrease rapidly. December and January form the coldest part of year. In the cold season, in the wake of western disturbances passing across north India, cold waves affect the area and on such occasions the minimum temperature drops down to about 3°C or 4°C.

Humidity

During the southwest monsoon season the relative humidity remarkably increases at Pachmarhi. The relative

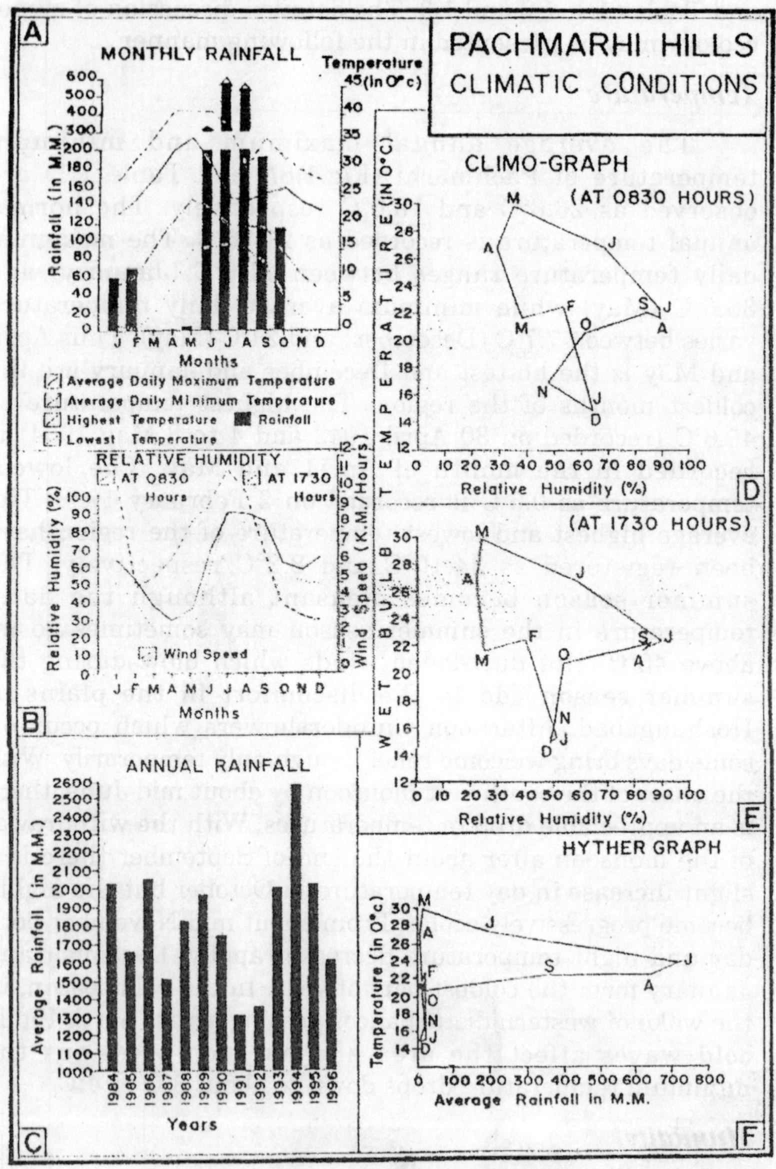
PACHMARHI HILLS
CLIMATIC CONDITIONS
A
MONTHLY RAINFALL
Temperature
45(in O°c)
Rainfall (in M.M.)
Months
J F M A M J J A S O N D
Average Daily Maximum Temperature
Average Daily Minimum Temperature
Highest Temperature
Rainfall
Lowest Temperature
B
RELATIVE HUMIDITY
AT 0830 Hours
AT 1730 Hours
Relative Humidity (%)
Wind Speed (K.M./Hours)
Wind Speed
Months
C
ANNUAL RAINFALL
Average Rainfall (in M.M.)
Years
1984 1985 1986 1987 1988 1989 1990 1991 1992 1993 1994 1995 1996
CLIMO-GRAPH
(AT 0830 HOURS)
TEMPERATURE (IN O°C)
Relative Humidity (%)
D
(AT 1730 HOURS)
WET BULB TEMPERATURE
Relative Humidity (%)
E
HYTHER GRAPH
Temperature (in O°c)
Average Rainfall in M.M.
F

Fig. 1.8

Table 1.1 Pachmarhi—Temperature (°C) and Relative Humidity (%)

Months	*Average Daily Maximum Temperature*	*Average Daily Minimum Temperature*	*Normal Temperature*	*Highest Temperature*	*Date of Recording*	*Lowest Temperature*	*Date of Recording*	*Relative Humidity At 0830 (IST)*	*Relative Humidity At 1730*
January	21.6	8.8	15.20	27.8	31 Jan 1964	1.1	16 Jan 1935	63	51
February	24.1	10.4	22.45	31.7	23 Feb 1953	0.6	02 Feb 1928	53	40
March	28.3	15.2	21.75	36.1	22 Mar 1982	23.3	01 Mar 1906	37	25
April	33.3	20.4	26.85	40.6	30 Apr 1942	8.9	02 Apr 1905	29	22
May	35.6	24.0	29.80	40.6	4-8 May 1954	15.0	10 May 1933	35	24
June	31.1	22.3	26.70	40.6	02 June 1889	15.6	30 June 1931	68	58
July	24.6	19.9	22.25	37.4	25 July 1960	16.1	15 July 1941	89	90
August	23.8	19.4	21.60	30.0	11 Aug 1899	15.0	17 Aug 1939	89	83
September	25.2	19.0	22.10	35.6	28 Sept 1931	12.8	30 Sept 1940	84	82
October	26 .3	14.9	20.60	31.7	04 Oct 1920	6.7	31 Oct 1933	61	54
November	23 .6	10.3	16.95	29.3	16 Nov 1957	2.2	30 Nov 1912	57	51
December	21.9	7.7	14.80	27.8	11 Dec 1941	1.1	27 Dec 1926	64	49
Average Annual	26.6	16.6	21.76	34.10		8.2		61	52

Source: District Gazetteers, Hoshangabad.

humidity percentage as measured at 0830 and 1730 hours are given in Table 1.1 and cartographically shown in Fig.1.8B. The humidity decreases in the post-monsoon season, and in the rest of the year the atmosphere is generally very dry. The driest part of the year is the summer season when the relative humidity is measured between 20 and 30 per cent. Table 1.1 indicates that the relative humidity (at 0830 hours) varies between 29 per cent (April) and 89 per cent (July and August). At 1730 hours, the relative humidity is recorded varying between 22 per cent (April) and 90 per cent (July) in the region. The average annual percentage of relative humidity, at 0830 and 1730 hours is recorded as 61 per cent and 52 per cent respectively.

Rainfall

The rainfall data of the region is given in Table 1.2 and it is further depicted by bar diagrams in Figs.1.8A and 1.8C. About 92 per cent of the annual rainfall is received in the southwest monsoon season. July being the rainiest month. The variation in the annual rainfall from year to year is appreciable. From 1984 to 1996, the highest annual rainfall as 2567.80 mm is recorded in 1994 while it is found lowest as 1300.20 mm in 1991. The average annual rainfall between 22 years is also measured as 1810.57 mm. The average annual rainfall of Pachmarhi is recently found as 1596 mm in 1996. It is measured maximum in July (578 mm) and August (540.80 mm) respectively. Pachmarhi experiences, on an average 80 rainy days in a year.

Cloudiness

During the southwest monsoon months skies are mostly heavily clouded of overcast. During the rest of the year the skies are generally clear or lightly clouded.

Winds

Winds are generally light. During the southwest monsoon season, winds are mainly southwesterly or westerly. In the post-monsoon and cold seasons winds are from the northeast or east. Southwesterlies and westerlies start in

March and these become predominant by May. Fig.1.8B and Table 1.2 show the wind speed (km/hours) of the region. The maximum and minimum wind speed have been recorded as 11.3 km/hours in the month of July and 2.9 km/hours in the month of December respectively. The average annual wind speed of the region is calculated as 6.9 km/hours in 1996.

Table 1.2 Pachmarhi—Average Rainfall and Wind Speed

Years	*Average Rainfall (in mm)*	*Months (1996)*	*Average Rainfall (in mm)*	*Wind Speed (km/hours)*
1984	2019.00	January	59.60	10.6
1985	1653.70	February	70.00	4.8
1986	2045.00	March	Nil	5.3
1987	1500.20	April	0.60	6.4
1988	1818.00	May	4.40	8.2
1989	1956.90	June	316.40	9.5
1990	1734.70	July	578.00	11.3
1991	1300.20	August	540.80	10.1
1992	1347.20	September	191.60	6.9
1993	1991.10	October	114.20	3.7
1994	2567.80	November	Nil	3.1
1995	2007.60	December	Nil	2.9
1996	1596.0			
Average	1810.57		1596.00	6.9

Source: District Gazetteers Hoshangabad.

Special Weather Phenomena

Depressions originating in Bay of Bengal in the Southwest monsoon season cross the east coast of India and more in some westerly direction. Such depressions pass through the area or its neighbourhood and cause widespread heavy rain and gusty winds. Occasional storms and depressions in the post-monsoon season also affect the weather of the region. In the summer season occasional dust storms and dust raising winds occur. Thunder storms occur in the summer season, and to a lesser extent in the cold season. Rain during the monsoon is also often associated with thunder. Fogs occur occasionally during the cold season, specially at the hill station. Table 1.3 depicts the events of special weather phenomena wherein thunder and fogs are the common elements experienced in the area. The climograph (Figs.1.8D and 1.8E) and hythergraph (Fig.1.8F) also indicate the climatic conditions of the area.

Table 1.3 Pachmarhi—Special Weather Phenomena

Weather	*Months (Showing average number of days)*												
Elements	*J*	*F*	*M*	*A*	*M*	*J*	*J*	*A*	*S*	*O*	*N*	*D*	*Annual*
Thunder	1.3	2.4	1.6	2.3	0.7	2.9	1.2	1.2	2 3	1.0	0.2	-	17.1
Hail	0.0	0.0	-	-	-	-	-	-	-	-	-	-	
Dust storm	0.0	0.0	-	-	-	-	-	-	-	-	-	-	-
Squall	0.0	0.0	-	-	-	-	-	-	-	-	-	-	-
Fog	0.4	0.0	-	-	-	3 .1	15.2	13.6	6.1	-	0.6	0.1	39.9

Source: District Gazetteers, Hoshangabad.

Pedology

Pedogenic environment can be studied by soil characteristics or pedology of the region. The pedology of the region is principally governed with various environmental factors like parent material, time, topography, climate, vegetation and living organisms. The above factors have changed mainly the composition and fertility of the soil in a number of ways from one place to other.

On the basis of the extensive field study and consulting several agricultural officials, the soils of the region have been grouped into the following categories:

(i) The upland soil

(ii) Riverine soil

The upland soil may be grouped as black soil, gravelly soil and sandy soil while the riverine soil may be explained as sandy soil in the upland region and alluvial soil in the lower part of the valley. The black soil occupies almost the whole region. It varies in depth and is usually loamy to clayey in texture. Lime concretion zone and free calcium carbonate are invariably present at different depths. Cracks develop in summer season and in deep clayey soil they are even a metre or more deep. This soil is usually ill supplied with phosphate, nitrogen and organic matter but is sufficient in potash and lime and suitable for cotton, jowar, wheat, sugarcane, groundnut etc.

So far as chemical composition of black cotton soil is concerned, 84 per cent soils indicate high percentage of potassium, 54.5 per cent are high in phosphorus while 81 per cent soils are low to medium in nitrogen, phosphoric acid and organic matter but potash lime and iron contents are usually high. The nitrogen content of 'Regur' soil is very low (0.02 to 0.05%), while phosphoric acid (0.08 to 0.2%), potash (0.8 to 0.15%) and lime (1.0 to 7.7%) also vary. This soil has three sub-types: (i) Deep black soil (ii) Medium black soil and (iii) Shallow black soil.

Deep black soil covers the level and open part of Satpura range. This soil may be further subdivided as black, dark brown, coarse brown, mixed and sandy. The highly productive black soils in this part are known as 'Mariar' and 'Kabar I'. They have clay content of 50-60 per cent and calcium carbonate percentage of 0.45. The pH ranges from 7.0 to 7.5.

Morand soil of Hoshangabad and Pachmarhi area is more or less medium black soil and is defined by the presence of small limestone pebbles and has a deposition of calcium carbonate in the lower layers. It is more gritty and friable than 'Kabar'. Shallow black soil is generally seen over the major ridges and in the southwest region. It consists of shallow loams having 15-30 per cent clay. The important types are dark brown clay, loamy rice soil, black soil and poor light hilly soil. Skeletal or gravelly soil consists of stony uplands of the region. It usually grows inferior millets and oilseeds.

Sandy soils called 'Domatta' and 'Retari' are found along streams and on higher ground suitable for Kharif crops.

Alluvial soils occur only in river valleys. Denwa and Sonbhadra rivers generally show the blanketing of alluvial soil in their lower reaches where flood occurs in rainy months.

Flora and Fauna

Flora

According to the classification proposed by H.G. champion the bulk of the forests in the area fall under 4-a, tropical dry mixed deciduous forests. Some area falls under Group 3-a, tropical moist mixed deciduous type. A few patches of forests and leading over Pachmarhi hills e.g. Bori reserves may correctly be grouped under Group 2, tropical semi-evergreen forests and those on highlands of Pachmarhi under Group 4, sub-tropical wet hill type. This classification is based mainly on climatic and ground conditions. In these forests the Teak *(Tectona grandis Linn. F.)* is the principal species among the high trees, because of its extraordinary surviving capacity against fires, grazing and maltreatment. The expansion of teak in a large area necessitates a clear division of all the forests biotically into (i) Teak forests where the species exist from 20 to 100 per cent of the stock and (ii) Mixed forests, where other species occur in preponderance balanching the spread of teak. The distribution of Teak is also determined by the parent rocks from which the soil is derived. It is intimately associated with and thrives exclusively on the Deccan trap and the resultant black cotton soil. On Trap flows or intrusions Teak predominates often to the virtual exclusion of other tree species. It may also thrive on metamorphics and the conglomerates of all stages of the upper Gondwanas, though of lesser quality and in small proportions to other forest species than on Trap. Like the Trap, Teak also prefers well-drained alluvial soil. However, it definitely avoids all types of the Gondwana sandstones which are invariably under the mixed type of forests Fig.1.9).

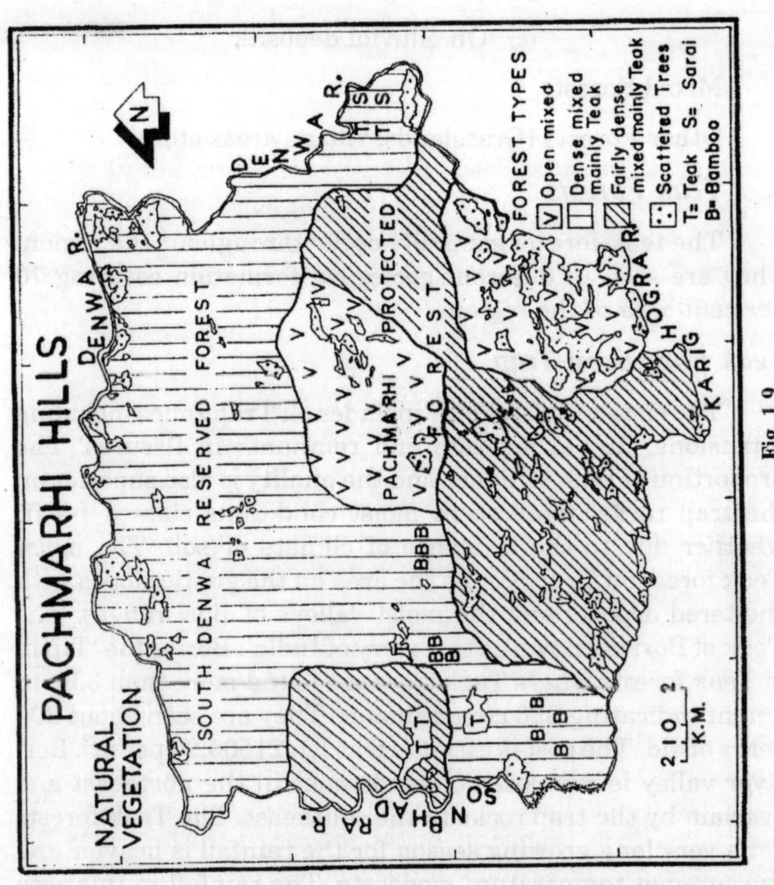

PACHMARHI HILLS
NATURAL VEGETATION
SOUTH DENWA RESERVED FOREST
PACHMARHI PROTECTED FOREST
DENWA R.
SONBHADRA R.
N
2 1 0 2
K.M.
FOREST TYPES
Open mixed
Dense mixed mainly Teak
Fairly dense mixed mainly Teak
Scattered Trees
T= Teak S= Sarai
B= Bamboo

Fig. 1.9

Thus the convenient ecological types and sub-types of forests are:

1. Teak Forests *(a)* On Trap *(i)* In moist conditions
   (ii) In dry conditions
   (b) On conglomerates and metamorphics
   (c) On alluvial deposits
2. Mixed Forests
3. Other Forests (Grasslands, village areas etc.)

1. Teak Forests

The teak forests generally occur throughout the region. They are seen in different geological formation covering 70 per cent area of the region.

Teak Forests on Trap

The class unmistakably includes all Trap zones including intrusions, fluvial deposits and remnants in fissures. The proportion of teak is higher and the quality is also superior on the trap rocks, more so in moist conditions classed I-a (i) whether due to the influence of climate or soil. The moist Teak forest on Trap occupy the area on the gentle slopes with sheltered aspects and the moist valleys of Bori ranges. The Teak of Bori is known as the 'glory of India'. Bori is the 'Tirth' of Teak forests where Teak trees are found more than 33m in height indicating 336 cm perimeter. They are seen about 200 years of old. The cost is estimated as Rs. 21500.00 per m^3. Bori river valley is composed of sandstones in the northeast and overlain by the trap rocks in the southeast. The Teak forests get a very long growing season for the rainfall is heavier and the summer temperature moderate. The rainfall in this area amounts to about 2500 mm. The Teak trees remain green in favourable localities upto April. The understorey is thick usually associated with Bamboo (Dendrocalamsu Strictus Nees). Such thickets invariably obstruct natural reproduction in which respect Dry forests are at an advantage. Generally the quality of Teak forests is approximately II (21 to 27 m) to III (15 to 21 m) in moist Teak forests with the height of the trees ranging from 27 m to 15 m.

Dry Teak forests, however are of lower quality and are open generally affected by Phuli grass *(Apluda varia),* Bhandar plant *(Strobilanthes callosus Nees)* and Makor *(A. Pennata)* and other climbers. Teak is the most precious wood for buildings, furniture, ships, steepers etc. Shisham *(Dalbergia Latifolia),* Ghiria *(Chloroxylon Swietenia DC),* Baranga *(Kydia Calycina),* Salal *(Boswellia Serrata Roxb)* and Bhandar *(Strobilanthes Collosus Nees)* are the trees associated with Teak forests.

Teak Forests on Conglomerates

Teak forests on the conglomerates and metamorphic rocks are generally poor in growth (30 cm to 60 cm in girth) as well as proportion (5 to 20%) with other species of mixed forests. They rarely attain the desired girth (90 cm) and become hollow at an early age. Such forests occur in Jatamau, Nilgarh, Bangra, Bineka, Khoramba Pipalgota, Satnal and Moran. In these tracts when Denwa conglomerates phyllites and schists, calcareous cystaline rocks and granitic gneisses are underlain by Bagra conglomerates, it increases to about 30 per cent. Where the Trap is mixed or underlying rock, it may attain high proportions upto pure stocks. The dry bed of Denwa conglomerate and the Pachmarhi sandstone generally show this type of forests. In Pachmarhi town, the Teak forests are seen in 169.970 sq km of area which is the 4.68 per cent area of the Hoshangabad district.

Teak Forests on Alluvial Strips

Teak forests of superior quality occur on the narrow alluvial strips along the Denwa from Brijikhapa to Mohari, the Sonbadhra to the north of Dhain and the Ganjal from Unchabari to Bothi and the Bori. These forests either discontinue or change suddenly in to the Mixed Forests wherever the Gondwana rocks crop out. The quality of the crop varies from II (height 15 to 21m) on the deep, fine moist and well drained soil to IV(a) (height 12 to 15 m) on mixed or sandy alluvium. The proportion of teak in the overwood varies from 20 to 80 per cent, the average being 50 per cent. The most familier features of this type are the occurrence of Bel (*Aegle marmelos*), the big trees sacred to Lord Shiva and Saj (*T. Tementosa*) and occasionally Reunjha (*Acacia Leucophloea*). Bamboos are often absent but the climbers especially Rauni (*Zizyphus Oenoplia*) are heavy on kachhars.

Mixed Forests

Mixed forests are the original type of forests in the region. Even the extensive teak forests are the results of the expansion of one single species under the adverse circumstances suppressing the rest of the associates. Thus, all the non-teak forests are classed as mixed forests. The mixed forests are not limited in the Gondwana formations. They are deciduous forests but remain green over a longer period than the nearby teak forests. The mixed forests occur about 35 per cent area of the region. The mixed forests show the main species of commercial importance are Saja and Dhaora.

Common Species of High Forests

Among the associate trees Ghiria is common everywhere on the sandy soil. Tendu is a small ebony tree commonly found on sandy loam. Ebony fruits are eaten. Mahua flowers attract both men and treasts alike. Bears and deer species are seen feeding on it before dawn in winter. Anjan is common on the Denwas and in the east of Sonbhadra river. Rohan is a common tree on calcareous soil; Bija or Bijasal is a common tree produces a good timber, and also used as medicine for diabetics. Salai is a general associate of teak on dry and stoney' localities and also seen along Denwa river. Other common species are Dhaora, Lendia, Achar, Semal, Moyan Haldu, Shisham, Kusum, Rinjha, Palas, Bel, Jamrasi, Mokha, and Phansi.

Useful Trees and Plants

Teak is the principal timber tree followed by Saj, Bija and Tinsa. Tendu Ghiria (*Chloroxylon Swietania*), Dhaura

(*Anogeissus Latifolia*) and other inferior wood are utilised as timber of low quality and are known as 'Jangal wood'. Siharee (*Nyctonthes arbortristis*), Ber (*Zizyphus nummularia*) and Bamboos are used for fencing. The best fuel wood are obtained from Dhaura, Khair (*Acacia Catechu*), Ber, Koha (*Terminalia arjuna*), lendia (*Lagerstroemta Parviflora*) and Kakai (*Flacourtia romantchi*).

Mahua, Achar, Mango, Tendu, Ber, Jamun, Niboo, Sitaphal *(Anona Squamosa)* are some of the fruit bearing trees. Roots of Pales *(Butea Frondosa),* Mahul *(Bauhinia Vahlii),* the bark of Anjan *(H. binnata)* and Babair grass are used for rope making.

Underwood

Bamboos so useful for matting and 'Tokni' making is also required on all religious occasions. They provide the necessary raw material for paper and synthetic industries. They are associated with Teak along the Bori valley. Papra trees are seen along the streams. Khair is a small tree which demands more light and avoids shade of high trees.

Rare Species

Divergence in the floral variety is specially noticeable around Pachmarhi where the occurrence of some species is a matter of Botanical interest. Sal *(Shorea robusta)* occurs in the catchment area of the Nagdwari, to the southwest of Pachmarhi and at a few other places. This is the western most limit of Sal in the Peninsula. According the report of the forest office of Pachmarhi, *Adintum Lyclosorus, Microsorium, Onicigonum, Lygodium Atherium, Asplenium, Tectoria, Chilanthus, Leucostenia, Gleichenia, Nephrolepis Tenifolia, Davallia, Polystichum, Etopteris, Osmunda, Blechnum, Pteridium, Cyatheae, Psitotuma, Alsophtla, Utricularia, Helianthus, Dryopteris, Lycopodium Ornnum Thalitrium foliolosum* and centella asiatica are the rare plant species found in the Pachmarhi region.

Climbers

The giant wood climber Mahul (*Bauhinia vahlii*) is very

heavy is sheltered moist localities, especially in the hills of Bori and Sohagpur ranges. Palas Bel (*Butea superba*) is found hanging on Teak and other associate trees. Gurar (*Milletia auriculata*) is very heavy in moist valley of Bori. Ironi is common in the overgrazed areas and mixed forests. Chilati (*Acacia Pennata*), Malkangani (*Celastras Paniculata*) Ramdaton (*Smilax macrophylta*) and Nagbel (*Cryptolapis buchanani*) are other notable climbers twin round the seedlings and poles. Lantana camara has obtained a foothold on Mankideo hill and certain river banks. Petalidium and Colebrookia are heavy in parts of Bori Bandar is dense under Teak. Gokhru (*Xanthium Strumarium*) and Tarota (*Cassia Tora*) colonizes the sites of cattle camps.

Grasses

Of the grasses Sabai (*Ischaemum anguistifolium*) is the most important. It occurs abundantly in sandy soils, cut up edges of nalas and the Bagra conglomerate and Denwa sandstone formations. Dub (*Cynodan dactylon*), Muchel (*Iseilema Wrighttii*), Keil (*Andropogon annulatus*) and Gunhari *(Anthistiria Scandnes)* are the best fodder grasses. Sukhal (*Pallinia argentea*) and Sen (*Ischaemum laxum*) are used far thatching village houses in the area. Babair grass (*Pallnia eriopoda*) exported to Kanpur and Poona for paper industry in the nineteenth century grows in plenty in the forests of Sohagpur tahsil. Now it is utilised locally for making ropes.

Apart from the above, the Pachmarhi special area shows various gardens developed by the forest administration in several localities (Table 1.4). The region shows 770 vegetation of 452 plant species under 101 families. The natural vegetation denotes 247 types of trees and 531 types of small leaf (Shak) plants.

Fauna

Owing to the plentiful supplies of food and water throughout the greater parts of the forests, the area used to be animal abode for wildlife. But during the past 50 years

Table 1.4 Gardens of Pachmarhi

Type of Garden	Name of the Garden	Established	Area (in Hectares)	Purpose
1. Gardens of fruits	1. Poloddyan (Polo Garden)	1948	10.800	Plants and fruits
	2. Pagara Garden	1907	15.800	Plants and fruits
2. Public Garden (well decorated)	3. Main Garden (Karbala Garden)	1907 1951-52	04.00	Public interest Specially for Tourist
	4. Rajendragiri Garden	1951-52	02.400	Office Garden
3. Garden at historical place	5. Pandav Cave Garden	1989	02.000	Historical and Tourist interest
4. Garden at Religious place	6. Jatashanker Garden	1989	0.900	Religious thrist
5. Garden for people of minority	7. Central Vidya Garden	1989	0.300	For minority peoples
6. Garden around the Administrative buildings	8. Rajbhawan Garden	1909-10	04.00	Governor Residence
	9. Champak Garden	1930-31	02.600	CM Residence
	10. Ravishanker Bhawan Garden	1981	04.500	CM Residence
7. Garden for Environmental protection	11. Satpura Garden	1985	02.000	Environmental Protection

Source: Udyan Karyalaya, Pachmarhi

the poaching has badly affected the animal population. The maximum damage was done by the persons having crop protection licence. Similarly intensive working of the forests has caused disturbance in the forests which has affected inbreeding of the hervivorus animals. The reduction in their number has also in turn affected the number of carnivorous animals. The carnivora has gone down terribly. The tigers and the panthers which could be seen with little effort in forests are now rare in occurrence.

The old District Gazetteer of Hoshangabad notes the absence of three important species viz. the elephant, the buffalo and the swamp deer. Tigers (*Felis rigris*) are seen in all the ranges but their occurrence is usually confined to the remote parts of extensive stretches of forests, preferably near the water. They are seldom found on very high land and plateaus. Bori range indicates the maximum population of Tigers. Leopards, Sloth bear, Sambhar, Indian Gazelle, Barking deer, Four horned Antelope, Blue Bull, Black Buck, Mouse Deer, Haena, Wild dog, Jungle cat, Wolf, Wild Bear, Fox, Jackal, Toiddy cat, Indian Ratel/Honey Badger, common mangose, Pangolin giant Squirrel, Porcupine, Momtorilizord, Crocodile, Pithon, Cobra, Red dogs, wild pigs, Bisons, are the main animals generally seen in the thick forests of the region.

As might be expected in a well watered area which in parts is forested and in parts cultivated bird life is abundant and varied. Common Pariah kite, Grey jungle fowl, Red jungle fowl, Red spur fowl, Peacock, Red Whiskered Bulbul, Night Jar, Koel, Pied, Cresled cuckoo, Paradise fly catcher, Jungle babbler, Myna, Blue throated Barbet, Parakeet, Green Pigeon, Green Bee-eater, Partridge Gly Quail, Dove, Wood Pecker, owl, vulture Bat, Tawny Eagle, Crested Serpant eagle, Horn Bill, Treeple, King fisher and Indian Robin are the major birds of the regions. In the streams are large number of fish species and also small cray-fish and crabs in the deeper holes. Even upto a great height one sees crocodiles in the rivers. The forest villagers catch large numbers of the bigger fish in the fish traps of numerous descriptions. Traps are set in the rivers when the fishes are migrating downstream at the end of the rains. Throughout the drier months of the year, they also kill the small fish in small pools by poisoning them with barks, leaves and seeds of various trees.

Sub-Surface Ecology and Groundwater Balance

A typical lava flow in parts of the region consists of a lower massive portion passing upwards into vesicular type with intra-trappean beds in the form of clays occurring between two flows. Primary opening occur in the form of vesicles or cooling joints. Secondary openings have been developed by fracturing due to tectonic effects as well as superficial phenomenon of weathering. Groundwater occurs in the interflow zones, both under table and confined conditions. Many of the wells drilled penetrate 2 to 4 lava flows and the occurrence of water is often at the contact of the first and second or second and the third flows. All aquifers at depths of more than 20 m below ground level have water under confined conditions with the massive units acting as aquicludes. Recharge to the aquifers takes place naturally through outcrops or fractures in confining massive trap units by leakage. Occurrence of trappean aquifer to a depth of about 80 m is certain. However wells trapping about 3 to 5 m of vesicular trappen unit below water table in peak summer month is easily capable of yielding more than 100 kilo litre per hour.

Crystalline rocks as a rule do not possess primary intergranular openings. Fresh crystalline rocks constituting granites and gneisses have of then less than one per cent porosity and negligible permeability. The water bearing properties of crystalline rocks are dependent upon the intensity of secondary opening due to jointing and weathering. The weathered zones are often followed by fractured zones with a varying thickness of 12 to 20 m. The occurrence and movement of groundwater in crystalline rocks is through weathered and fractured rocks. In the upper portion water moves through permeable rocks and in the lower portion through fissures and joints. The average yield of borewell to a depth of 80 m can be taken to be 2000 to 3000 litre per hour. Large diameter open wells in such formations yield about 75 to 90 kilo litres per hour.

For hydrogeological discussions, Gondwanas are considered as semi-consolidated formations. Groundwater occurs in them under both water table and confined conditions in the granular zones, comprising sandstones, grits and occasional conglomerates. Large diameter open wells within reasonable depths can yield 100 Klph, while tubewells drilled upto maximum depth of 150 m, tapping at least 50 m of cumulative thickness of good granular zones, are likely to yield at least 600 Lph for moderate draw-downs. More consolidated Gondwanas may have their yields of about 480 Lph also for moderate drawdowns.

Sinha (1995) reported that Gondwana formations are best water bearing horizons and can serve as the only good patch for irrigation by tubewells. Of course local patches of alluvium and weathered pockets of granitic formations also play an important role in groundwater exploitation by means of tubewells and dug wells.

Percolation of water and oozing out at a lower level on the bed of the rivers has been described earlier. Springs appearing on the rock cuts and foothills are numerous in the region. Around Pachmarhi one may mark perennial springs near the circuit house, Rorighat, Mahadeo caves, Kiwarkar Pahar, end Mankideo Pahar. These springs provide the large quantity of water supply. The water is found cold and sweet. Some notable examples of 'Hot Springs' have been reported at Anhoni and Anhoni Samoni. Anhoni is known as Badi Anhoni and located in Chhindwara district falling in toposheet 55 J/10. Here vigorous gas activity accompanies the water discharge. The spring emerges from the Intersection of ENE-WSW trending dolerite dyke cutting across coarse sandstone of Denwa stage of Gondwana Sequence and a NNW-SSE trending fault (marked F1-F1 on Fig.1.5A). The discharge of the hot spring is 60 litres per minute with a temperature of 56°C during January to May whereafter it decreases to 50 litres per minute with increase in temperature to 57.5°C during June. It is possible that the local groundwater which is mixing with the hot water depletes and 57.5°C results in decrease of discharge and increase in temperature during summer. No alteration is seen around the hot spring or along the channel through which the hot water flows (Thushu and Prakash, 1987).

The another site denoting hot spring at Anhoni Samoni is also known as Chhoti Anhoni locally. The spring is located in Hoshangabad district falling in toposheet 55 J/6 (Fig.1.5A). The surface manifestation of the activity consists of hot water Spouts enclosed by cemented tanks and Seepages along nala occurring over a distance of 300 m. Talchir siltstone which are traversed by coarse grained dolerite dykes near the hot spring. The cumulative discharge of the hot spouts is about 10 litres per minute with intermittent gas bubble activity at a temperature of 44°C whereas the total discharge of the seepage of' 40°C. Water in the nala is around 50 litres per minute. The periodic measurement of discharge between January and June did not reveal any change in the discharge nor in the temperature of the hot waters. No hydro-thermal deposit has been noticed around the hot water spout or along the seepages in the nala. The above two hot springs are noticed in the west and east portion out side of the area. The groundwater found in the area is cold and iron mixed. It creates gastric trouble and acidity improves after drinking the water. Occasional fever and headache are not noticed by using the water. Leucoderma becomes dominant due to the available water. Griping pain and diarrhoea are also common.

CULTURAL ECOSYSTEM

The modern history of Pachmarhi does not go far behind. In 1862 captain Forsyth J. reported about the Sylvan Charms of the place to the Chief-Commissioner (Sir Rechard Temple) of Hoshangabad. He was so much impressed that the captain was asked to go to Pachmarhi again to examine the possibility of using the plateau for establishing a sanitarium. On receiving the positive report a sanitarium was built which is now turned into family quarters for military personnel. However, the first building that was constructed was named Bison lodge which is now a centre of attraction for tourists as it is museum called "VANIKI SANGRAHALAYA". The total forested area of Pachmarhi was under the rule of "KORKU JAGIRDAR". It is reported that a military centre was established here in 1870 and after seven years a local body government, the Cantonment board functioning. Pachmarhi was too precious to the British. They did not permit general public to see it. Even the mates and coolies who were employed for construction work at Pachmarhi, were instructed to keep their colony on the foothills presently called Matkuli. The English did not want the town to grow into a city of affluence destroying the calm and repose so the rules were framed in such a way that construction of big houses or two-storey buildings, higher than the prescribed limit, was not possible. Even the Nawab of Bhopal who used to come here frequently to play POLO was not permitted to construct a Palace in the area of the cantonment and he could get the Palace built on the outskirts of the township. The Palace was later purchased by the rulers of Nasirgarh. The ruins can still be seen at the doorship of Pachmarhi (Bansal, 1991).

The Pachmarhi cantonment area was inhabited by the Indians who were serving as the British officers of the Army or other Gora Sahibs. In course of the time the cantt. developed into a town in 1931 and now according to the census of 1991, it has the total population as 10083 including 6063 males and 4020 females (Table 1.5) in 9.80 sq. kms areal extent. The other part of Pachmarhi is under the administration of the Special Area Development Authority (SADA) which was established in 1976. The SADA covers the geographical area of 49.21 sq.km including 520 residential houses and 530 households. The total population of SADA area (Pachmarhi Notified Municipality is 2412 wherein male and female population have been noticed as 1305 and 1107 according to 1991 census. According to Table 1.5 the total

Table 1.5 Census Abstract of Pachmarhi (1991)

Sl. No.	*Details of Census*	*Pachmarhi Notified Municipality*	*Pachmarhi Cantt*	*Pachmarhi Total Urban Area*
1.	Area in sq. kms	49.21	9.80	59.01
2.	No. of occupied Residential houses	520	1792	2312
3.	No. of households	530	1808	2338
4.	Total Population (including Institutional and Houseless Population)	2412	10083	12495
	(a) Male Population	1305	6063	7368
	(b) Female Population	1107	4020	5127
5.	Total Population in the age group (0-6)	412	1357	1769
	(a) Male Population	209	703	912
	(b) Female Population	203	654	857
6.	Scheduled Castes (Total)	325	1994	2319
	(a) Male Population	171	1041	1212
	(b) Female Population	154	953	1107
7.	Scheduled Tribes (Total)	548	700	1248
	(a) Male Population	272	339	611
	(b) Female Population	276	361	637
8.	Literates (Total)	1542	7598	9140
	(a) Male Population	942	5030	5972
	(b) Female Population	600	2568	3168
	(c) Scheduled castes literates (Total)	201	1296	1497
	(d) Male S.C. literates	121	763	884
	(e) Female S.C. literates	80	533	613
	(f) Total Scheduled tribes literates	188	381	569
	(g) Male S.T. literates	120	190	310
	(h) Female S.T. literates	68	191	259

Source: State/District Primary Census Abstract 1991, NIC, M.P. State Unit, Betul, Chhindwara 6 Hoshangabad districts.

population of Pachmarhi urban area is found as 12495 which includes 7368 males and 5127 females respectively. The total number of occupied residential houses and households are also reported as 2312 and 2338 respectively. The total population in the age group of 0-6 is registered as 1769 including 912 males and 857 females. Scheduled castes and scheduled tribes population have been marked as 2319 (1212 males and 1107 females) and 1248 (611 males and 637 females) respectively. The literates population is noticed as 9140 (5972 males and 3168 females) including 1497 scheduled castes (884 males and 613 females) and 569 Scheduled Tribes (310 males and 259 females). The total population corresponding to Scheduled Castes, scheduled tribes, literates, scheduled castes literates and scheduled tribes literate in Pachmarhi notified municipality and Pachmarhi cantt areas have been found as 325 and 1994, 548 and 700, 1542 and 7598, 201 and 1296 and 188 and 381 respectively. Pachmarhi town is the major urban area of the region. Except urban land and corresponding tourists centres, the entire hilly country is thickly forested and scarcely populated by scheduled tribes settlement. All the tourist centres are more or less well connected with Pachmarhi urban land by Pacca road.

There are 06 primary schools, 02 Junior high schools, 02 high school and 01 inter college and degree college in the region. About 75 per cent population is engaged in government services while 25 per cent population is engaged in business. Most of the population is migrated from associated seven districts (Narsinghpur, Sehore, Betul, Khandawa, Raisen, Dewas and Chhindwara districts) and from different localities of Maharashtra, Uttar Pradesh and Rajasthan provinces. It is located in Pipariya tahsil of Hoshangabad district and can easily be reach here approached by 54 kms road distance from Pipariya. Although the air distance between Pipariya and Pachmarhi is about 33.6 kms. It is also known as summer capital city of M.P. government. There are about 45 hotels and 25 restaurant. M.P. tourism and SADA also provide hotel facilities. The nearest airport is Bhopal (195 km) which is connected by regular flights with Delhi, Gwalior, Indore and Bombay. By rail one can reach here from Pipariya located on the Bombay-Howrah main line via Allahabad. Pachmarhi is well connected by regular bus services with Bhopal, Hoshangabad Nagpur, Pipariya and Chhindwara. M.P. tourism also operates regular coach services between Bhopal and Pachmarhi. Taxies are available for local communication. M.P. Tourism Development Corporation operates 7 accommodation units in Pachmarhi which cater to the requirements of all income groups. While Rock-End Manor (A heritage property) and Satpura Retreat offer luxury; Amaltas Sahakar Bunglow, Panchvati Cottages and Panchvati Huts offer good comfort. Holiday Homes is for those on a budget.

Training college and centre is also functioning in Pachmarhi. The colleges is an autonomous body meant exclusively for army personnel. It imparts education upto Master level in Education and other Military science subjects. The Brigadier is the commanding officer of the Army establishment. Though this town is not on the map of All India Tourism till recently but many great Indians had stayed here. The first and foremost comes the name of Dr. Rajendra Prasad, the First President of free India. He was so fascinated by the beauty of the place that he visited Pachmarhi twice and stayed here for several days. Dr. Lohia, Shri Jai Prakash Nerayan, J.B. Kriplani, Ashok Mehta are the few names who had taken abode here for recouping their lost energy. The first commander-in-chief General Cariappa had also worked here as an officer in the Army. The Wizard of Hockey the late Major Dhyanchand had served as a N.C.O. here.

People from film industry, though are very much fascinated by the untouched beautiful scenery and would like to shoot as many films as possible but the time involved in the journey to and fro makes it rather difficult for them to come here with all the paraphernalia essential for shooting a film besides bearing the expenses that would occur on the transport. Nontheless, a few films viz. *Quismat, Pal Do Pal Ka Sath, Massey Sahib, Thoda sa Rumani Ho Jayen and Electric Moon* (A film for channel 4 of the B.B.C.) have been filmed here.

The National Centre of Scouting and Guiding movement, the Bharat Scouts and Guides is also located here. Maharishi Mahesh Yogi has got a Peethum established here and Ved Vigyan Vishwa Vidya Peethum is being run. A training centre for T.M. for ladies is also run by the Peethum Authorities. The total income and expenditure of Pachmarhi SADA and Cantt areas are estimated (1991-92.) as Rs. 211.06 and Rs. 206.75 lakhs and Rs. 34.3 and Rs. 33.77 lakhs respectively. During the session of 1991-92, the total expenditures in SADA and Cantt areas of Pachmarhi in reference of Public works, Road light and Water supply have been estimated as Rs. 0.60 and Rs. 4.2, Rs. 1.00 and Rs. 76.3 and Rs. 0.50 and Rs. 1.37 lakhs respectively. The expenditures on education and roads are also found as Rs. 3.34 lakhs and Rs. 0.84 lakhs in Pachmarhi Cantt area (Sinha, 1994).

Pachmarhi hill resort is a unique hill resort which has so many things that you not find in any other hill stations. Though this hill resort is being developed by the Government but the pace is very slow and therefore you might have lacked modern facilities that you could get at other hill resorts but very candidly spoken it can be said that the uniqueness about Pachmarhi lies in its naturalness and undiluted environment. You must have seen that modernities have interfered very little with the natural sports and nature is in its most natural form. It is hoped that the hill resort will make an opportunity to all people coming here as a tourist and for other needs by providing natural gift of all kinds. Fig. 1.10 illustrates the major cultural landscapes of Pachmarhi.

CONCLUSION

On the basis of the above discussion, it may be concluded that the Pachmarhi hill resort is a nice piece of

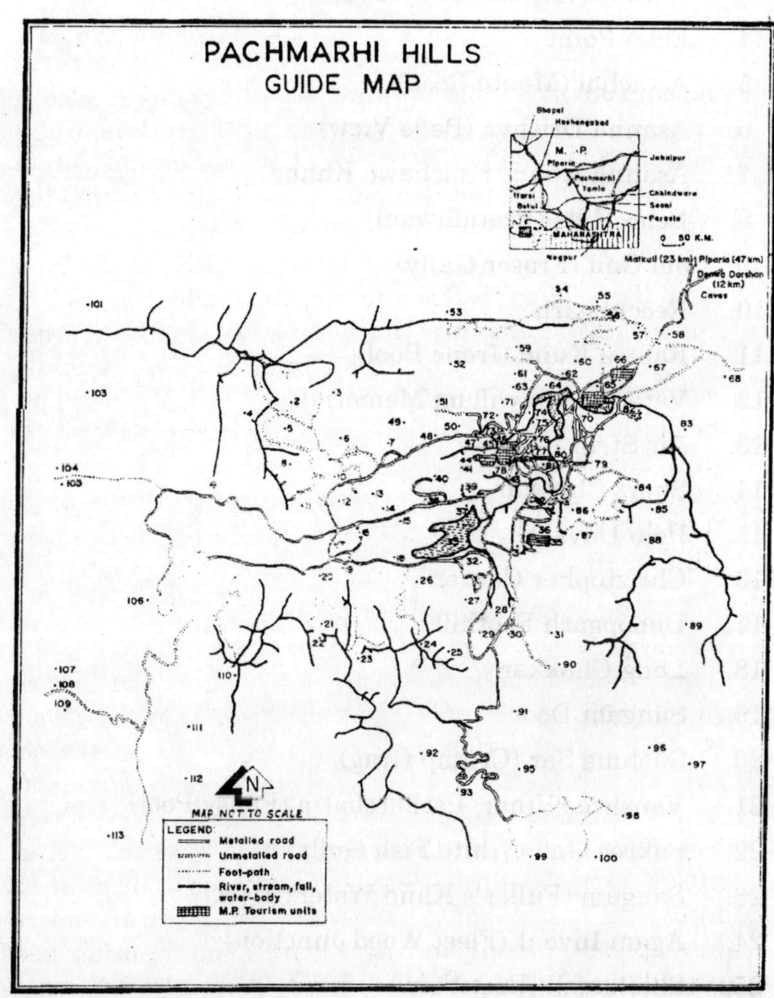
PACHMARHI HILLS
GUIDE MAP
Bhopal
Hoshangabad
M. P.
Jabalpur
Pipariа
Pachmarhi
Tamia
Itarsi
Betul
Chhindwara
Seoni
Parasia
MAHARASHTRA
Nagpur
0 50 K.M.
Matkuli (23 km)
Piparia (47 km)
Denwa Darshan (12 km)
Caves
MAP NOT TO SCALE
LEGEND:
Metalled road
Unmetalled road
Foot-path
River, stream, fall, water-body
M.P. Tourism units

Fig. 1.10

1. Sunderkund (Sa pool
2. Jalawataran (Duchess Hall)
3. Bhrant Neer (dorotiny Deep)
4. Echo Point
5. Astachal (Monte Rosa)
6. Asanna Drishya (Belle View)
7. Asanna Khud (Fanshawe Khund)
8. Echo Hills (Pratidhwani)
9. Jal Gali (Fraser Gully)
10. Reechhgarh
11. Ramya Kund (Irene Pool)
12. Vatsalya (Jwynham Memorial)
13. Air Strip
14. Short Chakkar
15. Polo Udyan
16. Christopher Corner
17. Dhoopgarh Foothill
18. Long Chakkar
19. Sangam Door
20. Sushma Sar (Crump Grag)
21. Vanshree Vihar, Patharchatta (Pansy Pool)
22. Sakree Gali (White Fish Gully)
23. Sangam (Fuller's Khud Waters Meet)
24. Agam Invent (Fleet Wood Junction)
25. Shailanjali (Best View)
26. Lat Shring (Lansdowne Point)
27. Rajendra Giri (Panorama Hill)
28. Ekant Giri (Mount Morris)
29. Long Khakkar

30. Handi Khoh
31. Handi Khoh Pahar (3,659)
32. Bharat Scouts and Guides
33. Padmini Jheel (Lake)
34. Rock Climbing Institute
35. Satpura Retreat
36. Carmel Mount
37. Champak Bangalow
38. Vanasthali Cottages
39. D.I.B. Amrak
40. Ravi Shankar Bhawan
41. Hog's Back (Vallabha Prastha)
42. New Hotel Block
43. Bison Public Lodge Gardens (Forest Museum)
44. Rock-Ena Manor
45. Neellamber
46. Panchvati Huts and Cottages China Bowl
47. PTO
48. Long and Short Chakkar
49. Lanji Giri (3,600)
50. Madhuban
51. Jamuna Prapat (Bee Fall)
52. Raj Giri (Club Hill)
53. Chhota Mahadev/Jambudwip Nala
54. Jambu Dwip
55. Brindavan
56. Brindavan
57. Nimbu Bhoj
58. Octroi Post

59. Twynham Pool
60. Bhainsa Sur Hill
61. Government Hospital
62. Civil Lines
63. Karnikar
64. Police Station
65. Cinema Bus Stand
66. Holiday Homes
67. Chusul (Army Education Crops)
68. To Kanjighat Village
69. Banganga Nala
70. Tourist Information Centre
71. Pachmarhi Club
72. Old Hotel Block
73. Raj Bhawan
74. Festival Ground
75. Christ Church
76. Suman Nandan Van
77. Youth Centre
78. Amaltas
79. Sahakar
80. Catholic Church
81. Prasthal
82. Pavas Prapat (Little Fall)
83. Sudrishya (Kitty Crag)
84. Poorva Darshan (Clematis Point)
85. Apsara Vihar (Fairy Pool)
86. Pandav Caves

87. Cemetery
88. Rajat Prapat (Silver Fall/Big Fall)
89. Bharna Pahar (2,772)
90. Chaura Darshan (Malcom Point)
91. Priyadarshini (Forsyth Point)
92. Mahdeo Pahar (4,360)
93. Bada Mahadeo
94. Denwa River
95. Gupta Mahadew
96. Bharna Village
97. Rocky Knob
98. Chauragarh (4,290)
99. To Nandiya
100. To Nandiya
101. Tamat Maidan
102. Nagdwari Nadi
103. Nandigarh Pahar
104. To Nagdwari Caves
105. To Kajri Village
106. Dhoopgarh (4,435)
107. To Kajri Village
108. To Bori Village
109. Ronghat Village
110. Tridhara (Picadilly Circus)
111. Reechhkhoh Pahar
112. Titanga Pahar
113. To Alimod Village

natural beauty. It has the richest sites of natural resources including biotic and abiotic forms. The assemblage of landforms and their complexity make the region more attractive and thus the attraction has caused to develop a new colony of happiness for tourists and the lovers of this lovely place.

REFERENCES

1. Ayyar, N.P., Verma, K.N., Nigam, R.K. and Singh R.Y., 1971: *Malwa Region, appeared in India, A Regional Geography,* edited by Singh R.L., pp.564-596.
2. Bansal, S.C., 1991: *Pachmarhi: A Trekkers Paradise Guide Book.*
3. Casshyap, S.M., 1977: *Pattern of Sedimentation in Gondwana Basins,* Fourth International Gondwana Symposium, G.S.I. Calcutta, 1, Hindustan Publishing Corporation Delhi, pp.1-34.
4. Casshyap, S.M., 1982: Palaeodrainage and Palaeogeography of Son-Valley, Gondwana Basin, M.P. Appeared in *Geology of Vindhyachal* edited by Valdiya and others, Hindustan Publishing Corporation, Delhi, pp.132-142.
5. Davis, W.M., 1900: The Geographical Cycle, *Geographical Journal,* Vol.14, pp.481-504.
6. Park, C.C., 1980: *Ecology and Environmental Management; A Geographical Perspective* (Butterworths) pp.109-110; 195-196.
7. Prasad, G., 1986: Environmental Modelling and Morphometric Analysis of Chitrakut Upland, Ph.D. Thesis Awarded by AMSE University, France.
8. Prasad, G., 1991: Morphological Processes, Land Forms and Their Significance in Environmental Management of Chitrakut and its Adjoining Region; Ph.D. Thesis submitted to Kanpur University, Kanpur.
9. Sinha, A.M., 1994: *Madhya Pradesh District Gazetteers, Hoshangabad,* Gazetteers Unit, Department of Culture, Government of M.P., Bhopal.
10. Sinha, A.M., 1995: *Madhya Pradesh District Gazetteers, Chhindwara,* Gazetteers Unit, Department of Culture, Government of M.P., Bhopal.

11 Spate, O.H.K., 1954: *India and Pakistan* (London, Methuen and Co. Ltd.), 7.

12. State/District Primary Census Abstract, 1991. N.I.C. U.P. State Unit, Hoshangabad, Betul and Chhindwara Districts, M.P.

13. Srivastava, P.N., 1990: Madhya Pradesh District Gazetteers, Betul District Gazetteers Unit, Department of Culture, Government of M.P., Bhopal.

14. Stoddart, D.L., 1950: Geography and Ecological Approach, the Ecosystem as a Geographical Principle and Method, *Geography*, Vol.50, pp.242-257.

15. Tansley, A.G., 1935: The Use and Abuse of Vegetational Concepts and Terms, *Ecology*, Vo1.16, pp.282-307.

16. Thassu, J.L. and Gyan Prakash, 1987: Assessment of Geothermal Potential of Hot Springs of Narmada Valley, *M.P. G.S.I. Records*, Vol.115, Part 6, pp.84-96.

17. Vildiya, K.S., 1982: "Tectonic Perspective of the Vindhyachal Region, Appeared in *Geology of Vindhyachal*, Edited by Valdiya and Others, Hindustan Publishing Corporation, Delhi, pp.23-29.

Topo-Ecology and Major Morphometric Determinants

In order to provide basis for the study and interpretation of the morphological processes and landforms of the region, a general framework of major morphometric properties including linear, areal and relief properties of drainage network, major components of drainage hierarchy, slope dynamics and drainage dissection have been introduced in this chapter. For this purpose 12 drainage basins of the region have been selected for critical analysis. The analysis and the results are mirrored through quantitative techniques based on topographical sheets.

LINEAR PROPERTIES OF DRAINAGE NETWORK

Stream Ordering

Stream order is a measure of a stream in the hierarchy of tributaries wherein the branching pattern of stream segments of all orders within a given watershed is analysed. Several geomorphologists like Horton (1932, 1945), Strahler (1964), Shreve (1966, 1967), Scheidegger (1965, 1966), Smart (1967, 1972) and Lewin (1970) have proposed the techniques of stream ordering (Fig. 2.1). Following the Strahler's stream segment method, the 12 drainage basins of the region have been segmented (Figs. 2.2 and 2.3) and the number of tributaries according to orders have been listed in Table 2.1. The total number of stream segments of different orders of 12 drainage

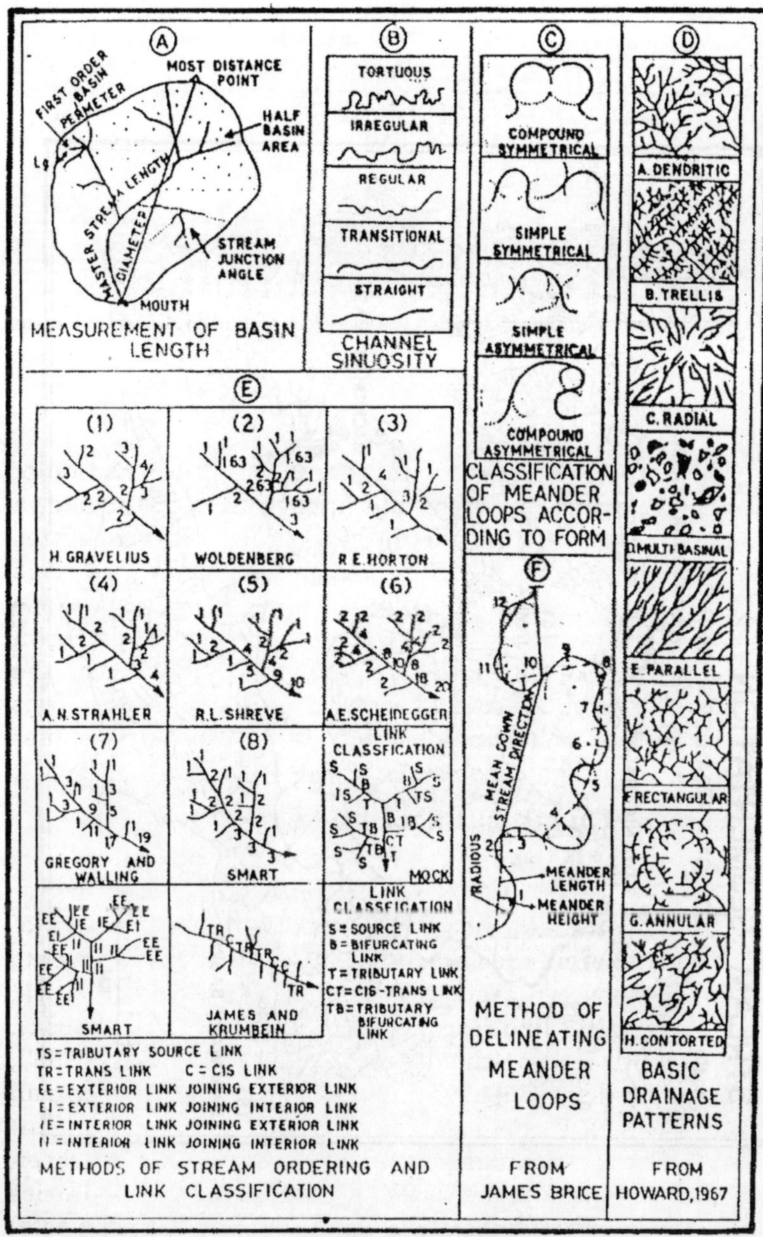
A
MOST DISTANCE POINT
FIRST ORDER BASIN PERIMETER
HALF BASIN AREA
Lg
MASTER STREAM LENGTH
DIAMETER
STREAM JUNCTION ANGLE
MOUTH
MEASUREMENT OF BASIN LENGTH
B
TORTUOUS
IRREGULAR
REGULAR
TRANSITIONAL
STRAIGHT
CHANNEL SINUOSITY
C
COMPOUND SYMMETRICAL
SIMPLE SYMMETRICAL
SIMPLE ASYMMETRICAL
COMPOUND ASYMMETRICAL
CLASSIFICATION OF MEANDER LOOPS ACCORDING TO FORM
D
A. DENDRITIC
B. TRELLIS
C. RADIAL
D. MULTI BASINAL
E. PARALLEL
F. RECTANGULAR
G. ANNULAR
H. CONTORTED
BASIC DRAINAGE PATTERNS
FROM HOWARD, 1967
E
(1) H. GRAVELIUS
(2) WOLDENBERG
(3) R.E. HORTON
(4) A.N. STRAHLER
(5) R.L. SHREVE
(6) A.E. SCHEIDEGGER
(7) GREGORY AND WALLING
(8) SMART
LINK CLASSFICATION
MOCK
SMART
JAMES AND KRUMBEIN
LINK CLASSIFICATION
S = SOURCE LINK
B = BIFURCATING LINK
T = TRIBUTARY LINK
CT = CIS-TRANS LINK
TB = TRIBUTARY BIFURCATING LINK
TS = TRIBUTARY SOURCE LINK
TR = TRANS LINK C = CIS LINK
EE = EXTERIOR LINK JOINING EXTERIOR LINK
EI = EXTERIOR LINK JOINING INTERIOR LINK
IE = INTERIOR LINK JOINING EXTERIOR LINK
II = INTERIOR LINK JOINING INTERIOR LINK
METHODS OF STREAM ORDERING AND LINK CLASSIFICATION
F
MEAN DOWN STREAM DIRECTION
RADIOUS
MEANDER LENGTH
MEANDER HEIGHT
METHOD OF DELINEATING MEANDER LOOPS
FROM JAMES BRICE

Fig. 2.1

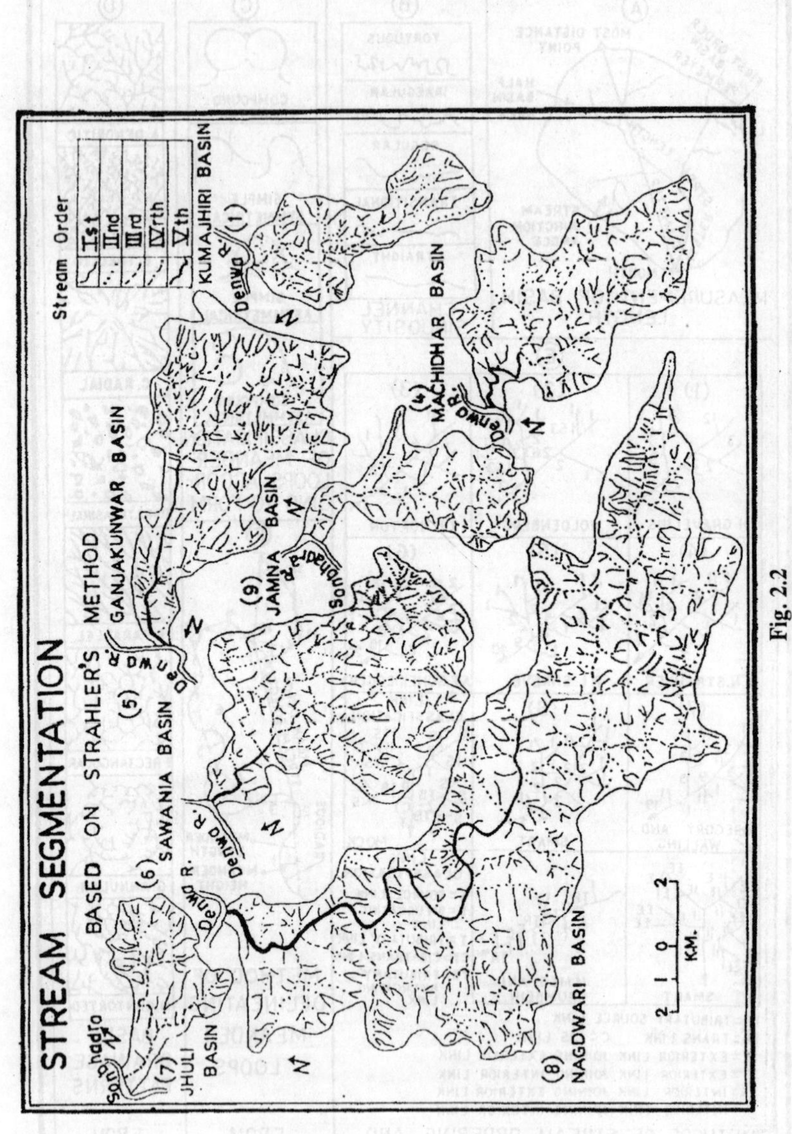
STREAM SEGMENTATION
BASED ON STRAHLER'S METHOD
Stream Order
Ist
IInd
IIIrd
IVrth
Vth
(1) KUMAJHIRI BASIN
(5) GANJAKUNWAR BASIN
(6) SAWANIA BASIN
(7) JHULI BASIN
(8) NAGDWARI BASIN
(9) JAMNA BASIN
(4) MACHIDHAR BASIN
Denwa R.
Sonbhadra
Sonbhadra R.
2 1 0 2
KM.

Fig. 2.2

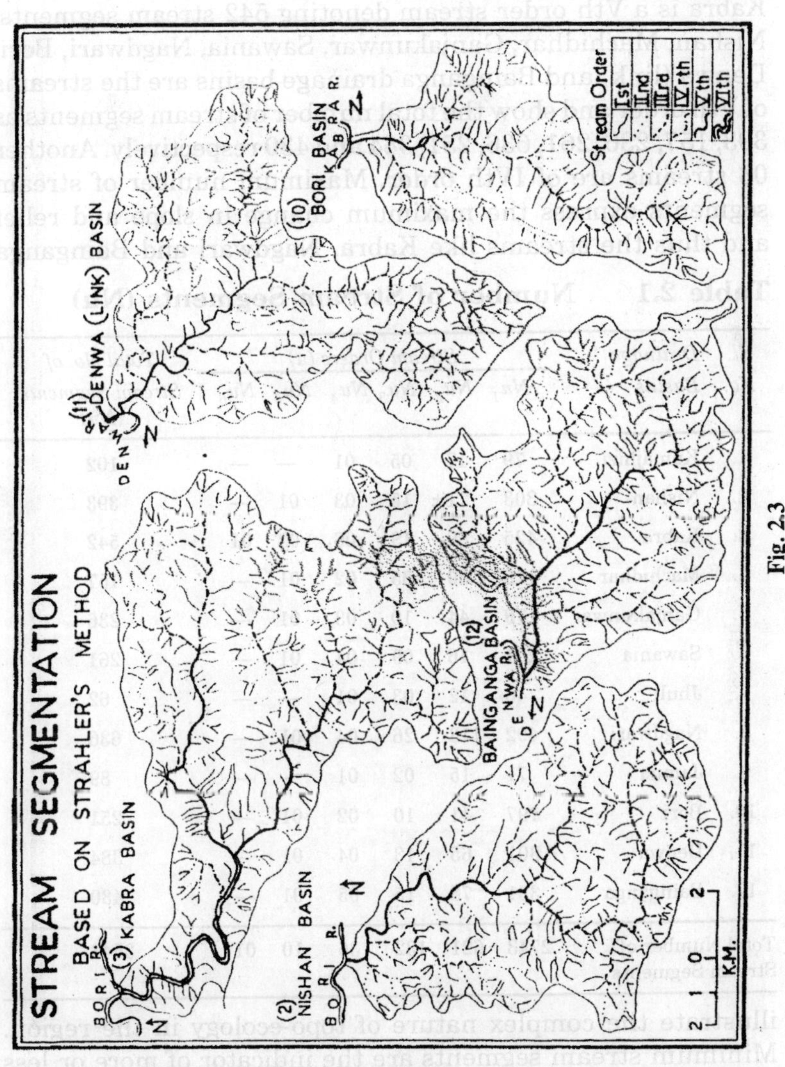
STREAM SEGMENTATION
BASED ON STRAHLER'S METHOD
(3) KABRA BASIN
BORI R.
(2) NISHAN BASIN
BORI R.
(11) DENWA (LINK) BASIN
DENWA R.
(10) BORI BASIN
KABRA R.
(12) BAINGANGA BASIN
DENWA R.
Stream Order
Ist
IInd
IIIrd
IVrth
Vth
VIth
2 1 0 2
K.M.
N

Fig. 2.3

basins have been calculated as 3543 (Table 2.1) wherein 2743, 621, 132, 36, 10 and 01 stream segments are noted against Ist, IInd, IIIrd, IVth, Vth and VIth orders of the basins respectively. Kabra is a Vth order stream denoting 542 stream segments. Nishan, Machidhar, Ganjakunwar, Sawania, Nagdwari, Bori, Denwa (link) and Bainganga drainage basins are the streams of Vth order and show the total number of stream segments as 393, 157, 236, 261, 636, 251, 384 and 430 respectively. Another 03 streams are of IVth order. Maximum number of stream segments express the maximum change in slope and relief and thus the streams like Kabra, Nagdwari and Bainganga

Table 2.1 Number of Stream Segments (Nu)

Sl. No.	*Drainage Basins*	*Stream Orders (u)*						*Total No. of Stream Segments (Nu)*
		Nu_1	Nu_2	Nu_3	Nu_4	Nu_5	Nu_6	
1.	Kumajhiri	79	17	05	01	—	—	102
2.	Nishan	303	71±	15	03	01	—	393
3.	Kabra	425	89	19	06	02	01	542
4.	Machidhar	119	30	05	02	01	—	157
5.	Ganjakunwar	178	44	10	03	01	—	236
6.	Sawania	201	46	09	04	01	—	261
7.	Jhuli	46	12	03	01	—	—	62
8.	Nagdwari	492	113	26	04	01	—	636
9.	Jamna	71	15	02	01	—	—	89
10.	Bori	197	41	10	02	01	—	251
11.	Denwa	301	65	13	04	01	—	384
12.	Bainganga	331	78	15	05	01	—	430
Total Number of Stream Segments		2743	621	132	36	10	01	3543

illustrate the complex nature of topo-ecology in the region. Minimum stream segments are the indicator of more or less flattish terrain having normal condition of relief and slope including drainage dissection. Drainage complexity also denotes the drainage dissection of the region.

Bifurcation Ratio (R_b)

Horton (1932) defined 'Bifurcation Ratio' as the relationship between the number of stream segments of a given order (Nu) to the number of stream segments of the next order which may be represented as:

$$R_b = \frac{Nu}{Nu + 1}$$

where, R_b = Bifurcation Ratio

Nu = No. of Stream Segments of a given order

and Nu+1 = No. of stream segments of the next higher orders

'Bifurcation Ratio' is a dimensionless property of the hierarchical system of a drainage basin. It is basically controlled by the physical eco-system (Prasad, 1986, 90).

The bifurcation ratio of 12 drainage basins has been calculated with the help of the Horton's formula. Mean bifurcation ratio of all the drainage basins (Table 2.2 and Fig. 2.4B) ranges between 3.53 (Kabra drainage basin) and 4.80 (Nagdwari drainage basin), Table 2.2 and Fig. 2.4A generally illustrate that the values of bifurcation ratios decrease with increasing orders. The following hypotheses regarding the nature and trend of bifurcation ratios have been examined.

Hypothesis—I

"Bifurcation ratios within a given region tend to decrease with increasing order because as order increases the percentage of streams that caolesce into a higher order tributary also increases ..... this increase is due to the diminishing amount of area available."

(Giusti and Schneider, 1965)

The above hypothesis holds good in the present analysis with exceptions. The hypothesis gives satisfactory results only in

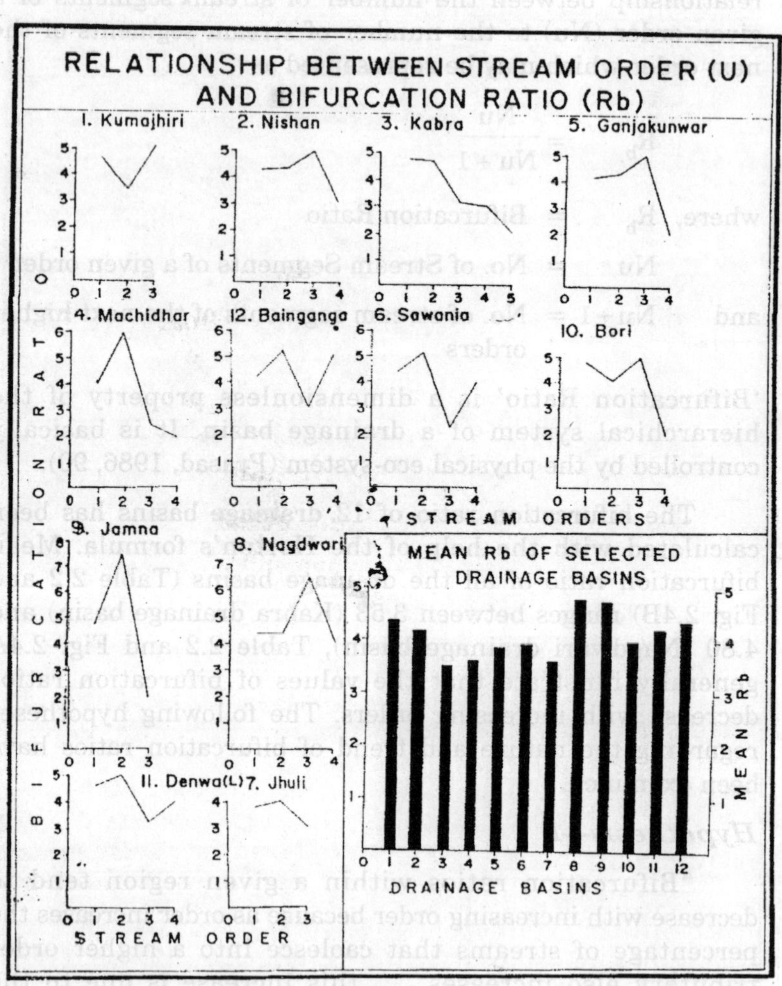

RELATIONSHIP BETWEEN STREAM ORDER (U) AND BIFURCATION RATIO (Rb)
1. Kumajhiri
2. Nishan
3. Kabra
5. Ganjakunwar
4. Machidhar
12. Bainganga
6. Sawania
10. Bori
9. Jamna
8. Nagdwari
11. Denwa(L)
7. Jhuli
BIFURCATION RATIO
STREAM ORDERS
STREAM ORDER
MEAN Rb OF SELECTED DRAINAGE BASINS
MEAN Rb
DRAINAGE BASINS

Fig. 2.4

Table 2.2 Bifurcation Ratio (R_b)

Sl. No.	*Drainage Basins*	$\frac{Nu_1}{Nu_2}$	$\frac{Nu_2}{Nu_3}$	$\frac{Nu_3}{Nu_4}$	$\frac{Nu_4}{Nu_5}$	$\frac{Nu_5}{Nu_6}$	R_b	*Area in sq.km.*
1.	Kumajhiri	4.65	3.40	5.00	—	—	4.35	15.00
2.	Nishan	4.27	4.37	5.00	3.00	—	4.16	70.00
3.	Kabra	4.78	4.68	3.17	3.00	2.00	3.53	99.00
4.	Machidhar	3.97	6.00	2.50	2.00	—	3.62	28.00
5.	Ganjakunwar	4.05	4.40	3.30	3.00	—	3.69	35.00
6.	Sawania	4.37	5.11	2.25	4.00	—	3.93	41.00
7.	Jhuli	3.83	4.00	3.00	—	—	3.61	11.00
8.	Nagdwari	4.35	4.35	6.50	4.00	—	4.80	115.00
9.	Jamna	4.73	7.50	2.00	—	—	4.74	15.00
10.	Bori	4.80	4.10	5.00	2.00	—	3.98	47.00
11.	Denwa (link)	4.63	5.00	3.25	4.00	—	4.22	73.00
12.	Bainganga	4.24	5.20	3.00	5.00	—	4.36	68.00

that circumstances where there is a smooth development of drainage basin under stable environmental conditions is existed but it is rare possible in nature.

Hypothesis—II

"The basin of equal order but variable areas tend to have the smallest R_b in the smallest area, the ratio increases with increasing areas upto a certain size beyond which R_b trends to become constant."

(Giusti and Schneider, 1965)

Following the Table 2.2 and Fig. 2.5 plotted against basin area and bifurcation ratios, it is pertinent to point out that the results of Rb and basin area in different orders have no definite trends. Fluctuations are marked in bifurcation ratios in respect of total basin areas. It is remarkable that the highest order (VIth order) basin like Kabra shows minimum Rb (3.53) with respect to the larger areas of 99 kms but IVth order basins like Jhuli (11 kms) and Vth order basins like Machidhar (28 kms^2), Ganjakunwar (35 km^2),

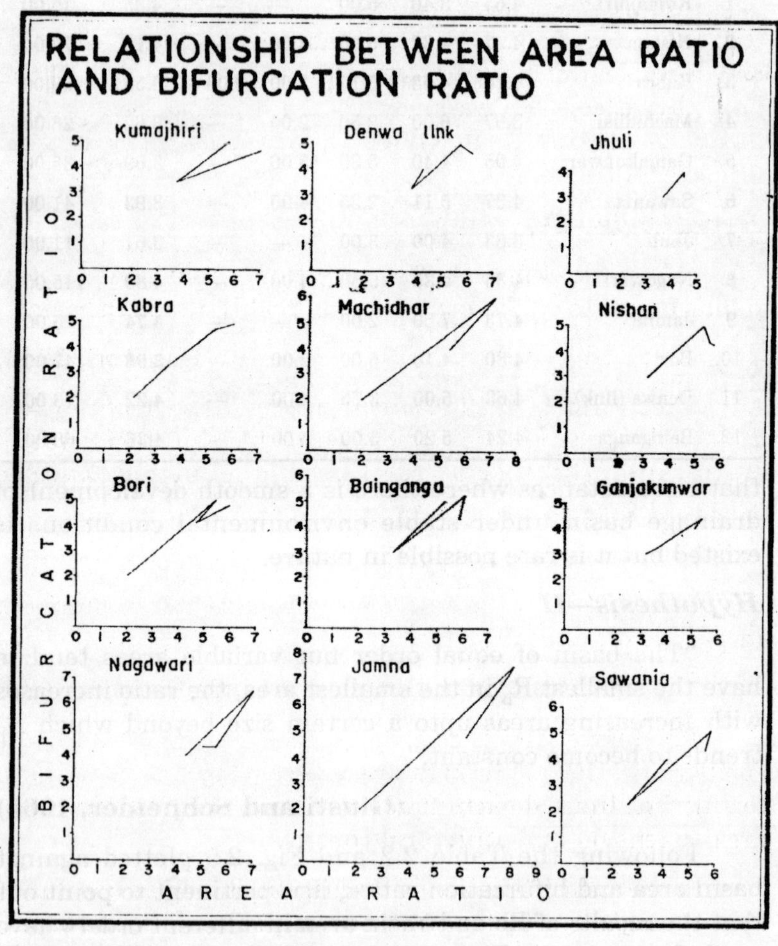
RELATIONSHIP BETWEEN AREA RATIO AND BIFURCATION RATIO
Kumajhiri
Denwa link
Jhuli
Kabra
Machidhar
Nishan
Bori
Bainganga
Ganjakunwar
Nagdwari
Jamna
Sawania
BIFURCATION RATIO
AREA RATIO

Fig. 2.5

Sawania (41 km^2) and Bori (47 km^2) show generally lowest R_b as 3.61, 3.62, 3.69, 3.93 and 3.98 respectively.

Hypothesis—III

"Bifurcation ratios tend to be affected by lithologies in that there may be more or less stable over uniform lithology and unstable over heterogeneous lithologies". It is remarkable here that except Kabra basin all the drainage basins show unstable marking of R_b which is caused by heterogeneous lithology of the region.

Hypothesis—IV

"Mean bifurcation ratios vary from about 20 for flat or rolling basins to 4.0 for mountainous, highly dissected basins."

(Horton, 1945)

The mean R_b of 12 drainage basins range between 3.53 and 4.80 which is the indicative of best performance of Hotanian law because all the basins drain highly dissected over hilly terrain.

Hypothesis—V

"Mean bifurcation ratios vary from region to region in respective of structural control."

(Miller, 1953; Singh, 1984 and Prasad, 1986)

The above hypothesis holds good results because the comparative study of R_b in different areas (Fig. 2.6) of northern foreland of India depicts much variation of R_b in respect of areal extension and structural controls.

In the light of the above discussion, it may be concluded that a single eco-factor may not controls a dimensionless property of mean bifurcation ratio but the several factors in association are responsible for it.

Stream Length

Stream length is a significant morphometric parameter which determines the drainage density and ruggedness of the

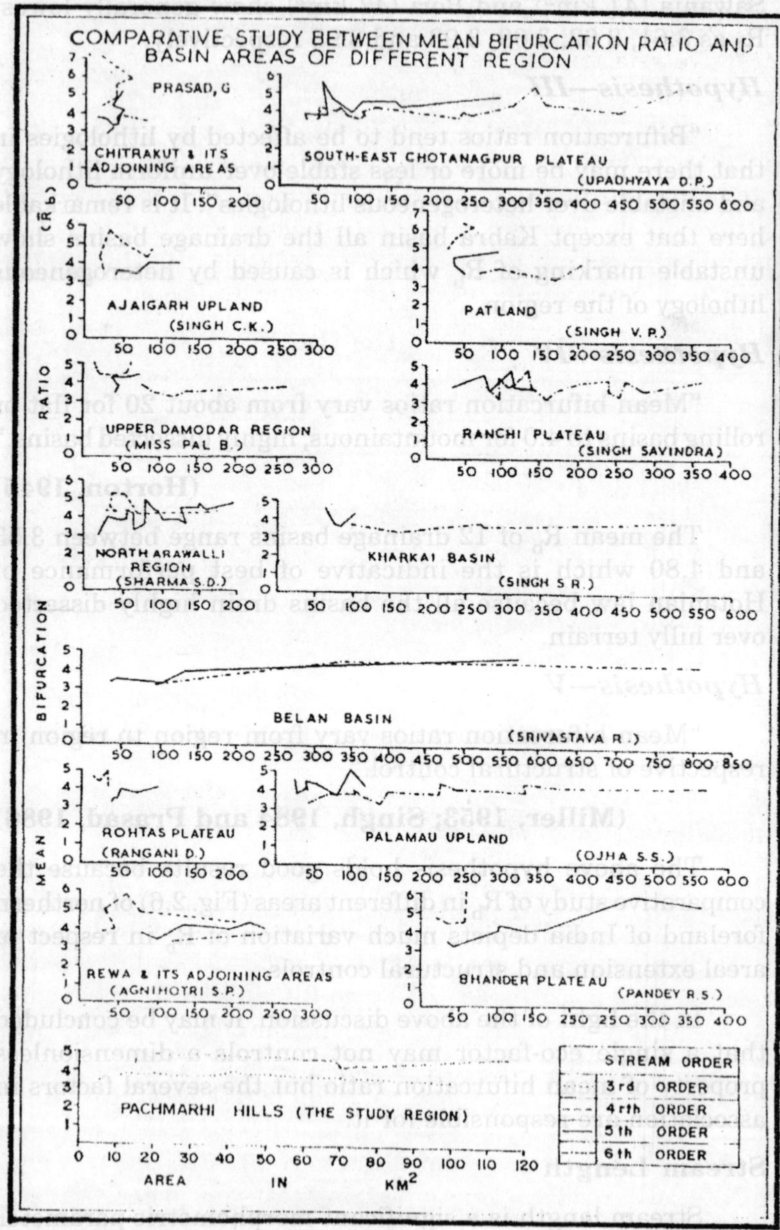
COMPARATIVE STUDY BETWEEN MEAN BIFURCATION RATIO AND BASIN AREAS OF DIFFERENT REGION
MEAN BIFURCATION RATIO ($\bar{R}_b$)
PRASAD, G
CHITRAKUT & ITS ADJOINING AREAS
SOUTH-EAST CHOTANAGPUR PLATEAU
(UPADHYAYA D.P.)
AJAIGARH UPLAND
(SINGH C.K.)
PAT LAND
(SINGH V. P.)
UPPER DAMODAR REGION
(MISS PAL B.)
RANCHI PLATEAU
(SINGH SAVINDRA)
NORTH ARAWALLI REGION
(SHARMA S.D.)
KHARKAI BASIN
(SINGH S.R.)
BELAN BASIN
(SRIVASTAVA R.)
ROHTAS PLATEAU
(RANGANI D.)
PALAMAU UPLAND
(OJHA S.S.)
REWA & ITS ADJOINING AREAS
(AGNIHOTRI S.P.)
BHANDER PLATEAU
(PANDEY R.S.)
PACHMARHI HILLS (THE STUDY REGION)
STREAM ORDER
3rd ORDER
4rth ORDER
5th ORDER
6th ORDER
AREA IN KM²

Fig 2.6

terrain. The stream length of each order in all the basins has been measured and arranged in Table 2.3. There are negative relationship between the total stream lengths and order. The total stream lengths of different orders of the basins tend to decrease proportionally with increasing orders over homogeneous topographic features and unstable under the heterogeneous topo conditions and controls. Basins like Nishan, Machidhar, Ganjakunwar, Jhuli, Dori and Bainganga exhibit decreasing trends of stream lengths with increasing orders. The remaining basins mark some deviations.

Table 2.3 Stream Lengths (Lu)

Sl. No.	*Drainage Basins*	*Stream Lengths (in kms)*					
		Lu_1	Lu_2	Lu_3	Lu_4	Lu_5	Lu_6
1.	Kumajhiri	33.50	15.00	3.00	6.00	—	—
2.	Nishan	129.50	37.00	21.00	15.50	8.00	—
3.	Kabra	181.00	43.00	19.00	16.00	16.50	8.50
4.	Machidhar	46.00	21.00	11.50	6.00	3.50	—
5.	Ganjakunwar	51.00	19.00	16.00	9.50	6.00	—
6.	Sawania	68.00	31.00	9.40	14.00	3.30	—
7.	Jhuli	18.00	5.50	4.25	4.10	—	—
8.	Nagdwari	198.00	74.00	34.00	14.50	21.00	—
9.	Jamna	31.00	10.50	4.50	4.70	—	—
10.	Bori	95.00	30.00	10.50	10.00	3.00	—
11.	Denwa (link)	140.00	41.50	15.00	24.50	8.00	—
12.	Bainganga	141.00	51 .00	15.50	15.00	8.30	—

The mean stream lengths of each order have been calculated and arranged in Table 2.4 and cartographically shown in Fig. 2.7. The average lengths of streams of each order in a drainage basin tend closely to approximate a direct geometric series wherein mean stream lengths increases with increasing orders. It is truly validated from Table 2.5 and Fig. 2.8. Deviations can be marked in some of the basins (Fig. 2.7) showing mean lengths against stream orders due to topographical controls.

Table 2.4 Mean Lengths (Lu) of Stream Segments (in kms)

Sl. No.	Drainage Basins	Lu_1	Lu_2	Lu_3	Lu_4	Lu_5	Lu_6
1.	Kumajhiri	0.42	0.88	0.60	6.00	—	—
2.	Nishan	0.43	0.52	1.40	5.17	8.00	—
3.	Kabra	0.43	0.48	1.00	2.67	8.25	8.50
4.	Machidhar	0.39	0.70	2.30	3.00	3.50	—
5.	Ganjakunwar	0.29	0.44	1.60	3.17	6.00	—
6.	Sawania	0.39	0.67	1.04	3.50	3.30	—
7.	Jhuli	0.39	0.46	1.42	4.10	—	—
8.	Nagdwari	0.40	0.65	1.31	3.62	21.00	—
9.	Jamna	0.44	0.70	2.25	4.70	—	—
10.	Bori	0.48	0.73	1.05	5.00	3.00	—
11.	Denwa (link)	0.47	0.64	1.15	6.13	8.00	—
12.	Bainganga	0.43	0 .65	1.03	3.00	8.30	—

Length Ratio (RL)

Length ratio (RL) is termed as the ratio between the mean lengths of stream segments of two successive orders and may be derived with the help of the following formula (Norton, 1945):

$$RL = \frac{\overline{Lu}}{\overline{Lu} - 1}$$

where

$$\overline{Lu} = \frac{Lu}{Nu}$$

where

$\overline{Lu}$ = Mean stream length of all segments of a given order

Lu = Total stream length of all segments of a given order, and

Nu = Number of stream segments of a given order.

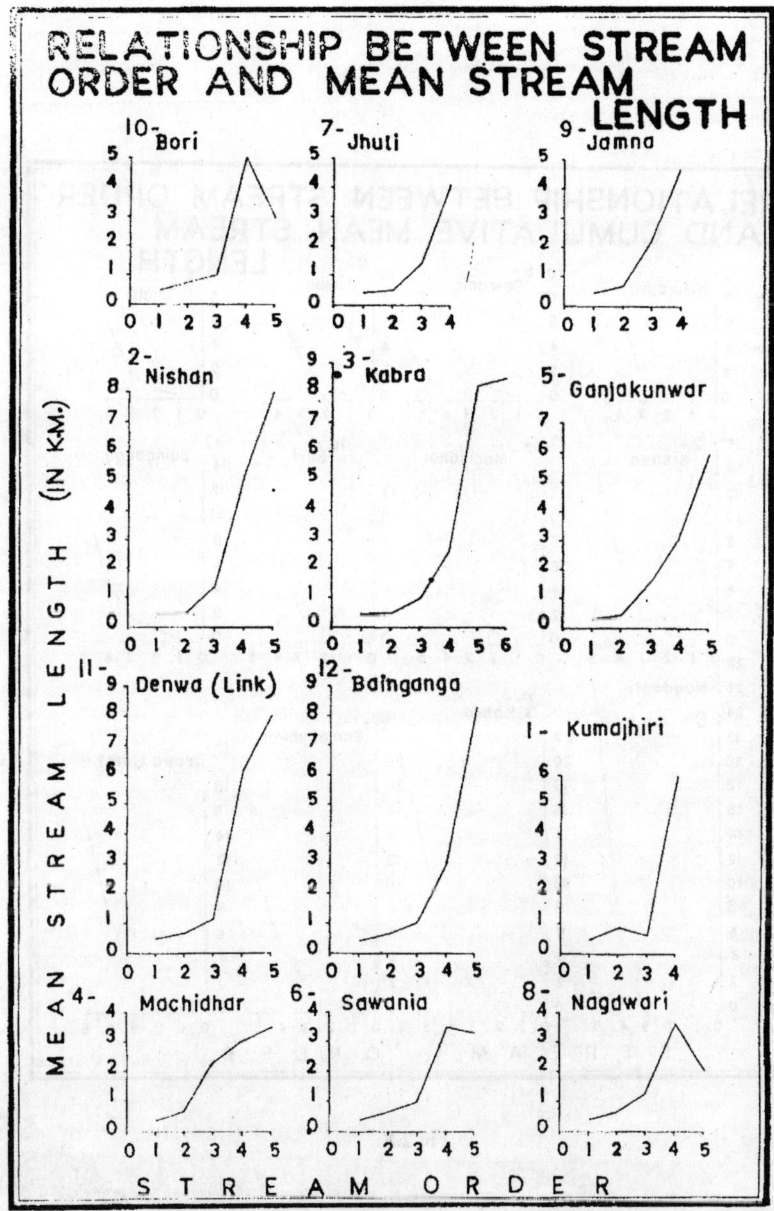
RELATIONSHIP BETWEEN STREAM ORDER AND MEAN STREAM LENGTH
10- Bori
7- Jhuti
9- Jamna
2- Nishan
3- Kabra
5- Ganjakunwar
11- Denwa (Link)
12- Bainganga
1- Kumajhiri
4- Machidhar
6- Sawania
8- Nagdwari
MEAN STREAM LENGTH (IN K.M.)
STREAM ORDER

Fig 2.7

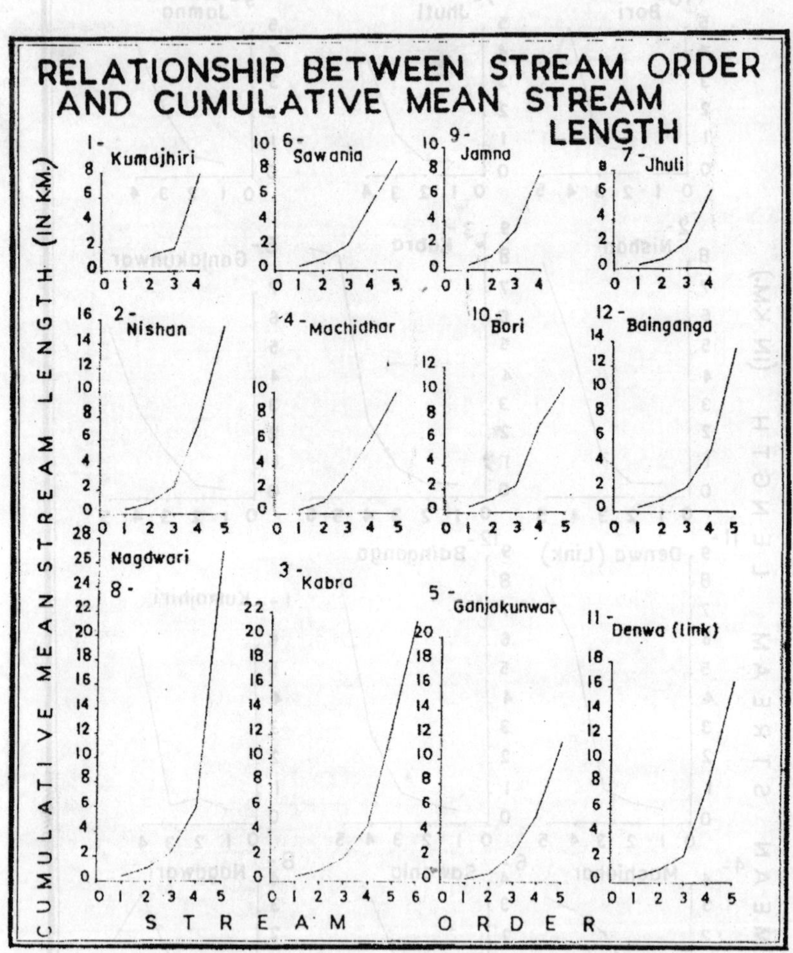
RELATIONSHIP BETWEEN STREAM ORDER AND CUMULATIVE MEAN STREAM LENGTH
CUMULATIVE MEAN STREAM LENGTH (IN KM.)
1- Kumajhiri
6- Sawania
9- Jamna
7- Jhuli
2- Nishan
4- Machidhar
10- Bori
12- Bainganga
8- Nagdwari
3- Kabra
5- Ganjakunwar
11- Denwa (link)
STREAM ORDER

Fig 2.8

Table 2.5 Cumulative Mean Lengths (in kms)

Sl. No.	*Drainage Basins*	Lu_1	Lu_2	Lu_3	Lu_4	Lu_5	Lu_6
1.	Kumajhiri	0.42	1.30	1.90	7.90	—	—
2.	Nishan	0.43	0.95	2.35	7.52	15.52	—
3.	Kabra	0.43	0.91	1.91	4.58	12.83	21.33
4.	Machidhar	0.39	1.09	3.39	6.39	9.89	—
5.	Ganjakunwar	0.29	0.73	2.33	5.50	11.50	—
6.	Sawania	0.39	1.06	2.10	5.60	8.90	—
7.	Jhuli	0.39	0.85	2.27	6.37	—	—
8.	Nagdwari	0.40	1.05	2.36	5.98	26.98	—
9.	Jamna	0.44	1.14	3.39	8.09	—	—
10.	Bori	0.48	1.21	2.26	7.26	10.26	—
11.	Denwa (link)	0.47	1.11	2.26	8.39	16.39	—
12.	Bainganga	0.43	1.08	2.11	5.11	13.41	—

The mean length ratio of drainage basins is shown in Table 2.6. It varies between 1.65 (Sawania drainage basin) and 4.26 (Kumajhiri drainage basin), significant variations of length ratio between one basin to others and even in a single basin are well marked. This wide range of variations are due to the topographical changes and lithological controls. Horton estimates that the values of the length ratios should range between 2 and 3. In the present case only seven (Nishan, Kabra, Machidhar, Sawania, Jhuli, Jamna and Bainganga drainage basins) drainage basins confirm the Hortanian law.

It is apparent from Table 2.6 that the highest order basins and older in age like Nishan (2.28), Kabra (2.06), Machidhar (1.82), Sawania (1.65), Nagdwari (3.05), Bori (2.08), Denwa link (2.45) and Bainganga (2.19) generally show lowest length ratios whereas the lowest order basins like Kumajhiri (4.26) and Ganjakunwar (6.96), comparatively younger in age exhibit highest mean length ratios.

On the basis of the above discussion, it may be inferred that younger the river, the more abnormal development of a stream length and also the higher abnormality in length ratios are observed. On the contarary, older the river, the more normal development of stream lengths and thus lesser abnormality in lengths ratios exist.

Table 2.6 Length Ratio (RL)

Sl. No.	Drainage Basins	$\frac{Lu_2}{Lu_1}$	$\frac{Lu_3}{Lu_2}$	$\frac{Lu_4}{Lu_3}$	$\frac{Lu_5}{Lu_4}$	$\frac{Lu_6}{Lu_5}$	RL
1.	Kumajhiri	2.09	0.68	10.00	—	—	4.26
2.	Nishan	1.21	2.69	3.69	1.55	—	2.28
3.	Kabra	1.12	2.38	2.67	3.09	1.03	2.06
4.	Machidhar	1.79	3.29	1.03	1.17	—	1.82
5.	Ganjakunwar	1.52	3.64	1.98	1.89	—	2.26
6.	Sawania	1.12	1.55	3.37	0.94	—	1.65
7.	Jhuli	1.18	3.09	2.89	—	—	2.29
8.	Nagdwari	1.63	2.02	2.76	5.80	—	3.05
9.	Jamna	1.59	3.21	2.09	—	—	2.29
10.	Bori	1.52	1.44	4.76	0.60	—	2.08
11.	Denwa (link)	1.36	1.79	5.33	1.31	—	2.45
12.	Bainganga	1.51	1.58	2.91	2.77	—	2.19

It is obvious that the length ratio between different orders of the basin do not follow any rule of either consistent decrease or increase with orders. As such, it becomes independent of drainage orders (Prasad 1987, 1990).

Sinuosity Indices

The condition in which the stream deviates alternately from adjacent to left and right walls of their most ideal expected flumes is expressed as a sinuous condition of a drainage basin (Prasad, 1987). Schumm (1963) has proposed five categories (Fig.2.1B) of channel sinuosity i.e. (i) straight, (ii) transitional, (iii) regular, (iv) irregular, and (v) tortuous. According to earlier findings, a stream having sinuosity of 1.0 is straight, just above the unity to 1.3 is sinuous and more than 1.3 is meandering. The following formulae as suggested by Schumm (1956) and Mueller (1968) have been used for the measurement of the sinuosity index of 12 drainage basins.

(i) $CI = \frac{C_L}{A_L}$ where CI = Channel Index

C_L = Channel Length

and A_L = Air Length between the source and the mouth of a river.

(ii) $VI = \frac{C_L}{V_L}$ where VI = Valley Index

and V_L = Valley Length

(iii) $S_{SI} = \frac{C_L}{V_L}$ where S_{SI} = Standard Sinuosity Index

(iv) H_{SI} = % equivalent of $\frac{C_I - V_I}{C_I - 1}$

where

H_{SI} = Hydrological Sinuosity Index or Percentage departure from a straight line course due to hydraulic factors

(v) T_{SI} = % equivalent of $\frac{V_I - 1}{C_I - 1}$

where

T_{SI} = Topographic Sinuosity Index or in other words percentage of stream departure from a straight course due to topographic interferences.

The measures of sinuosity indices have been tabulated in Table 2.7.

Table 2.7 Sinuosity Indices

Sl. No.	Drainage Basins	C_I	V_I	$H_{SI}(\%)$	$T_{SI}(\%)$	$S_{SI}(\%)$
1.	Kumajhiri	1.12	1.04	66.67	33.33	1.09
2.	Nishan	1.50	1.22	56.00	44.00	1.23
3.	Kabra	1.37	1.17	54.05	45.95	1.18
4.	Machidhar	1.41	1.29	29.27	70.73	1.09
5.	Ganjakunwar	1.32	1.16	50.00	50.00	1.14
6.	Sawania	1.38	1.22	42.11	57.89	1.13
7.	Jhuli	1.36	1.14	61.11	38.89	1.19
8.	Nagdwari	1.50	1.30	40.00	60.00	1.15
9.	Jamna	1.21	1.02	90.48	9.52	1.18
10.	Bori	1.19	1.04	78.94	21.06	1.15
11.	Denwa (Link).	1.55	1.33	44.00	66.00	1.16
12.	Bainganga	1.15	1.05	66.66	33.34	1.09

(a) Standard Sinuosity Index (S_{SI})

Table 2.7 carrying various sinuosity indices of 12 sample basins reveals the fact that there are not a single river which has its meandering course because the sinuosity index of all the basins range between 1.09 (Kumajhiri, Machidhar and Bainganga rivers) and 1.23 (Nishan river) which is the indicative of sinuous pattern of the rivers. The highland of Pachmarhi is much dissected and corresponding slopes are marked steep, so the region shows the straight flow path of the river due to structural controls.

(b) Hydraulic and Topographic Sinuosity

On the basis of the Table 2.7, it is clear that the basins like Kumajhiri (66.67%), Nishan (56.00%), Kabra (54.05%), Genjakunwar (50.00%), Jhuli (61.11%) Jamna (90.48%), Bori (78.94%) and Bainganga (66.66%) show highest H_{SI} percentage than that of T_{SI} percentage and denote the hydraulic dominance in the area. Drainage basins like Machidhar (70.73%), Sawania (57.89%), Nagdwari (60%) and

Denwa link (66%) exhibit the highest percentage of topographic sinuosity index and thus advocate the dominance of topography in their catchment areas. H_{SI} ranges between 29.27 per cent (Machidhar basin) and 90.48 per cent (Jamna basin) while T_{SI} varies between 9.52 per cent (Jamna basin) and 70.73 per cent (Machidhar basin) respectively.

APEAL PROPERTIES OF DRAINAGE NETWORK

Basin Perimeters

Basin perimeter is the linear measurement of a particular drainage net on the basis of water-shed line and treated as natural boundaries of the basins. It is related to the basin area. Table 2.8 reveals the fact that the basin area generally increases with increasing perimeters but the basins which are recognised as in elongated shape or highly controlled by associating lithology e.g. Kumajhiri and Jhuli register longer the lengths of basin perimeter with smaller basin areas.

Table 2.8 Different Basin Parameters

Sl. No.	*Drainage Basins*	*Basin Area (A) (in sq.km)*	*Basin length (VL) (in km)*	*Channel length (CL) (in km)*	*Air length (AL) (in km)*	*Basin Perimeters (in km)*
1.	Kumajhiri	15.00	7.80	8.50	7.50	19.50
2.	Nishan	70.00	15.30	18.80	12.50	37.00
3.	Kabra	99.00	20.50	24.10	17.50	55.00
4.	Machidhar	28.00	12.00	13.10	9.30	26.00
5.	Ganjakunwar	35.00	12.30	14.00	10.60	32.50
6.	Sawania	41.00	11.50	13.00	9.40	30.50
7.	Jhuli	11.00	6.30	7.50	5.50	15.50
8.	Nagdwari	15.00	28.65	33.00	22.00	67.50
9.	Jamna	15.00	7.35	8.70	7.20	20.50
10.	Bori	47.00	12.70	14 .60	12. 20	33.00
11.	Denwa (link)	73.00	17.35	20.10	13.00	46.00
12.	Bainganga	68.00	13.90	15.20	13.20	32.00

It is also apparent that the channel and basin lengths of 12 drainage basins increases with increasing length of basin perimeter. The channel lengths of all the basins found greater than that of the basin lengths. The smaller basins generally show the greater lengths of basin perimeter because the water dividing lines are much fluctuated by topographical change.

Basin Shape

The ideal drainage basin is usually of pear shape but since it is dominant on the size and length of the master stream of the basin and basin perimeter which are themselves dependent on other variables (Prasad, 1987). On an average, three sub-categories of basin shape have been recognised viz. circular, elongated and idented. Although various methods for the measurement of shape indices have been suggested but the following three methods are applied in the present study.

(i) Horton's (1932) Form Factor (F)

$$F = \frac{A}{L^2}$$

where A = Basin Area and

L = Basin Length

(ii) Mueller's (1953) Circularity Index (C)

$$C = \frac{4\Pi A}{P^2}$$

where P = Basin Perimeter

(iii) Chorley, Malm and Pogorzeiski (1957) Lemniscate Method (K)

$$K = \frac{L^2}{4A}$$

It is notable that the higher values of circularity index indicate the more circular shape of the basin while the lower values of 'F' and higher the values of 'K' indicate the more elongated shape of the basin. On the basis of the above formulae, the derived results of 'C', 'F' and 'K' have been arranged in Table 2.9. The low (below 30%), medium (30-60%) and high (above 60%) circularity index illustrate the youth, mature and old stages of drainage basins. Thus except Nishan and Bainganga all the basins show the mature stage of terrain development. The higher values of circularity index are the indicative of circular shaped basins. Form factor ranges between 0.14 (Nagdwari basin) and 0.35 (Bainganga basin) which show the more elongated shape of the basins. In the similar way, the lemniscate 'K' ranges between 0.71 (Bainganga basin) and 1.78 (Nagdwari basin) and the higher values of 'K' depict the elongation of the basins.

Table 2.9 Basin Shape Indices

Sl. No.	*Drainage Basins*	*Circularity Index (C)*	*Form Factor (F)*	*Lemniscate (K)*
1.	Kumajhiri	0.50	0.25	1.014
2.	Nishan	0.64	0.30	0.87
3.	Kabra	0.41	0.24	1.06
4.	Machidhar	0.52	0.19	1.29
5.	Ganjakunwar	0.42	0.23	1.08
6.	Sawania	0.55	0.31	0.81
7.	Jhuli	0.58	0.28	0.90
8.	Nagdwari	0.32	0.14	1.78
9.	Jamna	0.45	0.28	0.90
10.	Bori	0.54	0.29	0.86
11.	Denwa (link)	0.43	0.24	1.03
12.	Bainganga	0.83	0.35	0.71

Law of Basin Area

The basin area becomes cumulative from 1st order to the successively higher orders and thus the area of a higher order stream includes the total area covered by all the streams of lower order. There are positive relationship between stream order and basin area. The basin area increases with an increase in order of stream segments (Table 2.10). It is clear from Table 2.10 that the basins like Kabra, Nishan, Nagdwari, Denwa (link) and Bainganga generally show the greater basin areas than that of the others. In all the cases the basin area increases with increasing orders.

Table 2.10 Basin Areas (Au) in sq.kms.

Sl.No.	Drainage Basin	Au_1	Au_2	Au_3	Au_4	Au_5	Au_6
1.	Kumajhiri	08.00	12.00	13.00	15.00	—	—
2.	Nishan	38.50	53.50	62.50	66.50	70.00	—
3.	Kabra	54.00	70.00	80.00	86.00	92.00	99.00
4.	Machidhar	15.50	21.00	25.00	27.00	28.00	—
5.	Ganjakunwar	19.00	24.00	30.00	33.00	35.00	
6.	Sawania	25.00	31.00	35.00	40.00	41.00	—
7.	Jhuli	7.00	8.00	9.00	11.00	—	—
8.	Nagdwari	66.00	85.00	98.00	105.00	115.00	—
9.	Jamna	10.00	13.00	14.00	15.00	—	—
10.	Bori	29.00	37.00	42.00	46.00	47.00	—
11.	Denwa (link)	39.00	52.00	59.00	69.00	73.00	—
12.	Bainganga	38.00	50.00	58.00	64.00	68.00	—

Area Ratio (Ra)

The area ratio (Ra) is the proportion of increase of mean basin area ($\bar{A}u$) between two successive orders of a given basin. It is derived on the basis of the following formula (Strahler, 1969).

$$Ra = \frac{\bar{A}u}{\bar{A}u - 1}$$

where $\bar{A}u$ is the mean area of given order of a basin

$$\text{when } \bar{A}u = \frac{\bar{A}u}{Nu}$$

where Nu = Number of stream segments of a given order

and Au = The total area of all segments of the same order.

The mean basin area (Table 2.11) have been obtained and mean Ra of 12 sample basins have been also derived (Table 2.12).

Table 2.11 · Mean Basin Areas (Au) in Sq.kms.

Sl.No.	*Drainage Basins*	$\bar{Au}_1$	$\bar{Au}_2$	$\bar{Au}_3$	$\bar{Au}_4$	$\bar{Au}_5$	$\bar{Au}_6$
1.	Kumajhiri	0.10	0.70	2.60	15. 00	—	—
2.	Nishan	0.13	0.75	4.17	22.17	70.00	—
3.	Kabra	0.13	0.79	4.21	14.33	46.00	99.00
4.	Machidhar	0.13	0.70	5.00	13.50	28.00	—
5.	Ganjakunwar	0.11	0.56	3.00	11.00	35.00	—
6.	Sawania	0.12	0.67	3.89	10.00	41.00	—
7.	Jhuli	0.15	0.67	3.00	11.00	—	—
8.	Nagdwari	0.13	0.75	3.77	26.25	115.00	—
9.	Jamna	0.14	0.87	7.00	15.00	—	—
10.	Bori	().15	0.90	4.20	23.00	47.00	—
11.	Denwa (link)	0.13	0.80	4.54	17.25	73.00	—
12.	Bainganga	0.11	0.64	3.87	12.80	68.00	—

Table 2.12 Area Ratio (Ra)

Sl. No.	*Drainage Basins*	$\frac{Au_2}{Au_1}$	$\frac{Au_3}{Au_2}$	$\frac{Au_4}{Au_3}$	$\frac{Au_5}{Au_4}$	$\frac{Au_6}{Au_5}$	*Ra*
1.	Kumajhiri	7.00	3.71	5.77	—	—	5.49
2.	Nishan	5.77	5.56	5.32	3.16	—	4.95
3.	Kabra	6.08	5.33	3.40	3.21	2.15	4.03
4.	Machidhar	5.38	7.14	2.70	2.07	—	4.32
5.	Ganjakunwar	5.09	5.35	3.67	3.18	—	4.32
6.	Sawania	5.58	5.81	2..57	4.10	—	4.52

(Contd....)

Sl. No.	Drainage Basins	$\frac{Au_2}{Au_1}$	$\frac{Au_3}{Au_2}$	$\frac{Au_4}{Au_3}$	$\frac{Au_5}{Au_4}$	$\frac{Au_6}{Au_5}$	Ra
7.	Jhuli	4.47	4.48	3.67	—	—	4.21
8.	Nagdwari	5.77	5.03	6.96	4.38	—	5.54
9.	Jamna	6.21	8.05	2.14	—	—	5.47
10.	Bori	6.00	4.67	5.48	2.04	—	4.55
11.	Denwa (link)	6.15	5.68	3.80	4. 23	—	4.97
12.	Bainganga	5.82	6.05	3.31	5.31	—	5.12

Table 2.12 reveals the fact that mean Ra of 12 drainage basins varies between 4.03 (Kabra drainage basin) and 5.54 (Nagdwari drainage basin). Likewise mean length ratio, mean Ra varies among the basins of equal order but should decrease with the basins of increasing order. It is found that most of the rivers show decreasing trends of Ra with increasing stream orders. Fluctuations are marked due to some topographical and lithological changes. The basins like Kumajhiri (5.49) and Jamna (5.47) are of the lower order show the higher values of Ra.

RELIEF PROPERTIES OF DRAINAGE NETWORK

Relief Ratio (Rh)

Relief ratio is a dimensionless variable which denotes the ratio between height and basin length. It expresses the nature of the gradient of long profiles of drainage basins which in turn controls the intensity and magnitude of dissection mainly valley incision through denudational processes because it is the gradient of the longitudinal course of the channel which decides the nature of discharge, velocity of water flow and transporting capacity of the river concerned (Prasad, 1987). Schumm (1956) has defined the relief ratio as the ratio between the total relief of the basin (elevation difference of lowest and highest points of a basin) and the longest dimension of the basin parallel to the principal drainage line and thus presented the following formula for the calculation of relief ratio:

$$Rh = \frac{H - L}{CL}$$

where

Rh = Relief Ratio

H = Highest point of the basin at the source

L = Lowest point of the basin at the mouth

and CL = Channel length parallel to the principal drainage line.

The relief ratios of 12 drainage basins have been calculated and arranged in Table 2.13. The relief ratio ranges between 0.014 m/m (Nishan drainage basin) and 0.073 m/m (Jamna drainage basin). Relief ratio generally increases in respect to shorter channel length and higher maximum basin relief, over the highly dissected hilly terrain and decreases gradually according to longer channel length and the lower maximum basin relief over the flattish terrain.

Table 2.13 Relief Ratio and Relative Relief Percentage

Sl. No.	*Drainage Basins*	*Maximum Basin Relief in m (H)*	*Channel length (CL) in kms*	*Basin Perimeters (P) in km*	*Relief Ratio (Rh) m/m*	*Relative Relief Percentage (Rhp)*
1.	Kumajhiri	540	8.50	19.50	0.063	2.77
2.	Nishan	280	18.80	37.00	0.014	0.76
3.	Kabra	768	24.10	55.00	0.031	1.40
4.	Machidhar	587	13.10	26.00	0.044	2.26
5.	Ganjakunwnr	700	14.00	32.50	0.050	2.15
6.	Sawania	661	13.00	30.50	0.051	2.17
7.	Jhuli	420	7.50	15.50	0.056	2.71
8.	Nagdwari	992	33.00	67.50	0.030	1.47
9.	Jamna	640	8.70	20.50	0.073	3.12
10.	Bori	720	14.60	33.00	0.049	2.18
11.	Denwa (link)	782	20.10	46.00	0.038	1.70
12.	Bainganga	808	15.20	32.00	0.053	2.52

Relative Relief Percentage (Rhp)

Relative relief percentage is too much similar to relief ratio. Melton (1957) has designated his model of computation of relief ratio on the basis of the maximum basin relief and basin perimeter and termed relief ratio as relative relief percentage. The Rhp can be obtained with the help of the following equation:

$$\text{Rhp} = \frac{100\text{H}}{5280(\text{mile})/1000(\text{km})\,\text{P}}$$

where

Rhp = Relative relief percentage

H = Maximum basin relief

and P = Basin perimeter (in mile or km)

Table 2.13 represents the relative relief percentage of 12 drainage basins. The values of Rhp range between 0.76 (Nishan drainage basin) and 3.12 (Jamna drainage basin). Basins like Kumajhiri, Machidhar, Ganjakunwar, Sawania, Jhuli, Jamna, Bori and Bainganga generally show higher the values (more than 2) of Rhp while Nishan, Kabra, Nagdwari and Denwa (link) illustrate the lower values (below 2). Rhp increases as well as the length of basin perimeter decreases with higher maximum basin relief. It decreases according to increasing trends of basin perimeter having lower values of maximum basin relief. It shows similar findings as marked in Rh.

Relative Relief (RR)

Relative relief is a term used to refer a concept intended to describe the vertical irregularities of landscape. It is as the difference of heights between the highest and the lowest points in a given unit following to grid square method (Prasad, 1988).

On the basis of the grid square method, the relative relief has been obtained for the entire region and the total frequency and frequency percentage under four categories have been given in Table 2.14.

Table 2.14 Frequency Distribution of Relative Relief (RR)

Sl. No.	*Relative Relief categories (in m)*	*Taxonomic Explanation*	*Grid Frequency*	*Frequency Percentage*
1.	Below 60	Low relative relief (RL)	141	13.53
2.	60 - 120	Moderate relative relief (RM)	277	26.58
3.	120 - 180	High relative relief (RH)	271	26.01
4.	Above 180	Very high relative relief (RVH)	353	33.88
	Total		1042	100.00

Table 2.14 and Fig. 2.9 incorporate the areal distribution of relative relief in the region. The distribution recites the fact that there are symmetrical distribution of grid frequencies under various categories of relative relief as it is indicated by 13.53 per cent, 26.58 per cent, 26.01 per cent and 33.88 per cent under low, moderate, high and very high RR categories respectively. The Pachmarhi shows higher the values of relative relief which ranges between 04 m and 546 m. The grid frequency registered as 141, 277, 271 and 353 in low, moderate, high and very high categories clearly indicates that the region is much influenced by vertical irregularities. It is apparent from Table 2.14 that more than 50 per cent (59.89%) area of the entire region shows high and very high relative relief percentage. The major hills top like Khamkhera Pahar, Bariam Pahar, Bharna Pahar, Handikhoh Pahar, Chauragarh Pahar, Lathikhap Pahar, Kiwarkhap Pahar, Duranandiya Pahar, Mahadev Pahar, Richhkhoh Pahar, Dhupgarh Pahar, Pandidev Pahar, Barchachar Pahar, Nandigarh Pahar, Nishangarh, Barghat, Mahuldeo, Belkhandav, Ratomati, Tatragarh, and Mankidev Pahars and the associated valleys of Denwa, Bainganga, Jambudeep, Nagdwari, Bori and Kabra basins which show the highest peaks and lowest valley bottoms associated to each other confirm the high and very high variations of relative relief. Except few parts of valley floor

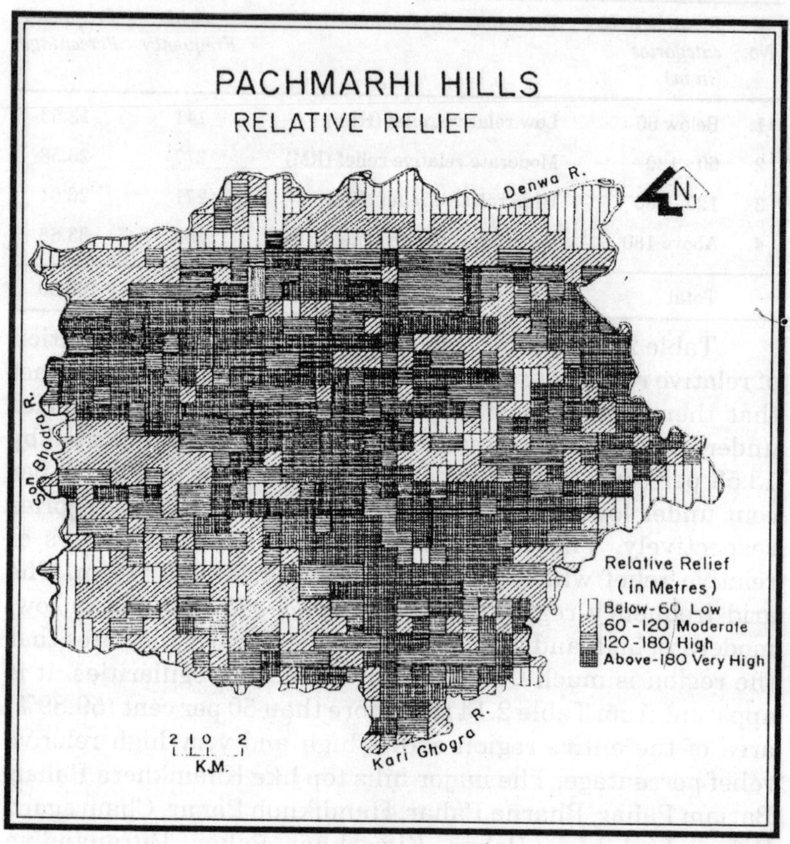
PACHMARHI HILLS
RELATIVE RELIEF
Denwa R.
N
Son Bhadra R.
Kari Ghogra R.
Relative Relief
(in Metres)
Below-60 Low
60-120 Moderate
120-180 High
Above-180 Very High
2 1 0 2
K.M.

Fig. 2.9

and the foothill zones, the flat top surfaces and the urban land of Pachmarhi have marked low to moderate relative relief. It is because of minimum relief variations at highest elevation. But the whole region is the land of high relative relief and the land of relief complexities.

Absolute Relief (AR)

Absolute relief is the highest elevation marked in a grid of one km^2/one mile^2 drawn on the topographical sheet. It is noted in each grid with the help of highest contour or relative height or Benchmark obtained in the grid area. Using grid square method, the absolute height or relief has been observed and it is further grouped in the following categories (Table 2.15).

Table 2.15 Frequency Distribution of Absolute Relief (AR)

Sl. No.	*AR categories (in m)*	*Taxonomic Explanation*	*Areal Frequency*	*Frequency Percentage*
1.	Below 4000	Moderate Absolute Relief (Hilly country) ARM	08	0.77
2.	400 - 800	Moderately High Absolute Relief (Mountainous country) ARMH	610	58.54
3.	800 - 1200	High Absolute Relief (ARH) (High Mountainous country)	404	38.77
4.	Above 1200	Very High Absolute Relief (ARVH) (Very High Mountainous country)	20	1.92
	Total		1042	100.00

According the Fig. 2.10 and Table 2.15, it is clear that very limited area (0.77%) below the height of 400 m is covered by moderate absolute relief while the mountainous country (above 400 m) is observed over 99.23 per cent of the total geographical area. Most of the extreme north, east, southeast and the southwest portion of the area have been marked by moderately high absolute relief. The middle portion including some highest peaks is demarcated by high absolute relief.

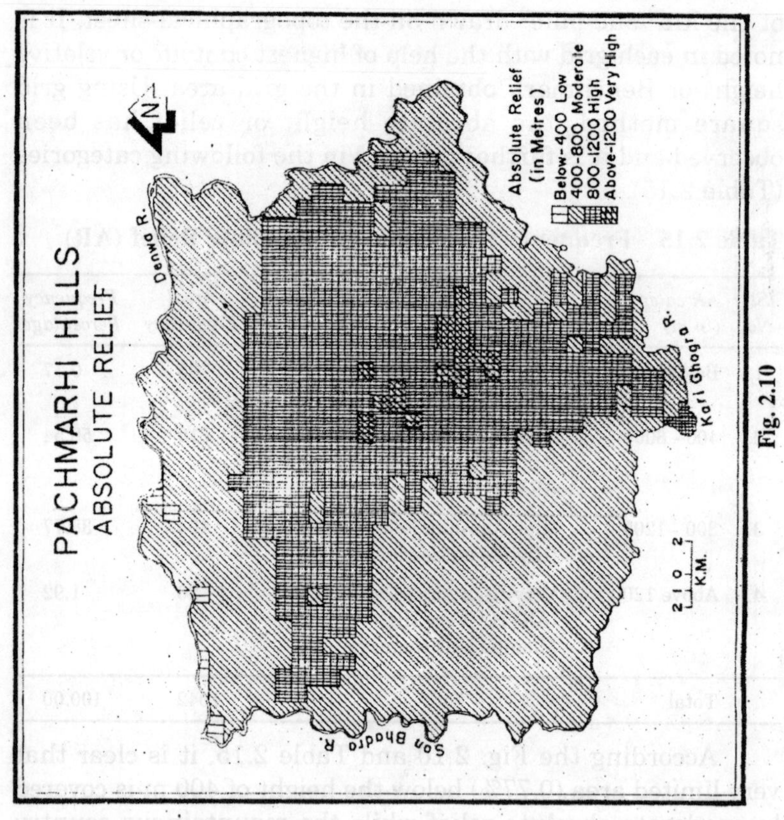
PACHMARHI HILLS
ABSOLUTE RELIEF
N
Denwa R.
Son Bhadra R.
Kali Ghogra R.
Absolute Relief
(in Metres)
Below-400 Low
400-800 Moderate
800-1200 High
Above-1200 Very High
2 1 0 2
K.M.

Fig. 2.10

MAJOR COMPONENTS OF DRAINAGE HIERARCHY

Drainage Density (Dd)

Drainage density has been appreciated as a fundamental indicator of the dynamic nature of drainage basin forms in a fluvially dominated terrain. It is a valuable index of drainage basin processes as it helps to reflect several environmental factors like climatic, topographic, lithologic, pedologic and vegetal and at a greater extent the influence of human interferences inflicted upon it. It is defined as the length of streams per square unit area (Prasad, 1987). Several geomorphologists like Horton (1932 and 1945), Carton and Langbein (1960), Gardiner (1971, 1974 and 1979), Mccoy (1970), Donahue (1972), Cariston (1963), Cotton (1964), Prasad (1987, 1989, 1990, 1993, 1995) etc. have studied this aspect of drainage basins.

The drainage density of the whole region has been derived using grid square method wherein grids are marked covering 1 km^2 area on the topographic maps (Scale 1:50,000). On the basis of Table 2.16 and Fig. 2.11 it may be explained that extremely low and low drainage density

Table 2.16 Frequency Distribution of Drainage Density (Dd)

Sl. No.	*Dd Categories*	*Taxonomic Explanation*	*Grid Frequency*	*Frequency Percentage*
1.B	Below 1	Extremely low drainage density (DdEL)	24	2.30
2.	1 - 2	Low drainage density (DdL)	137	13.15
3.	2 - 4	Moderate drainage density (DdM)	741	71.11
4.	4 - 6	Moderately high drainage density (DdMH)	138	13.25
5.	Above 6	High drainage density (DdH)	02	0.19

categories together represent 15.45 per cent grid frequency of the region. This type of features are shaded along the

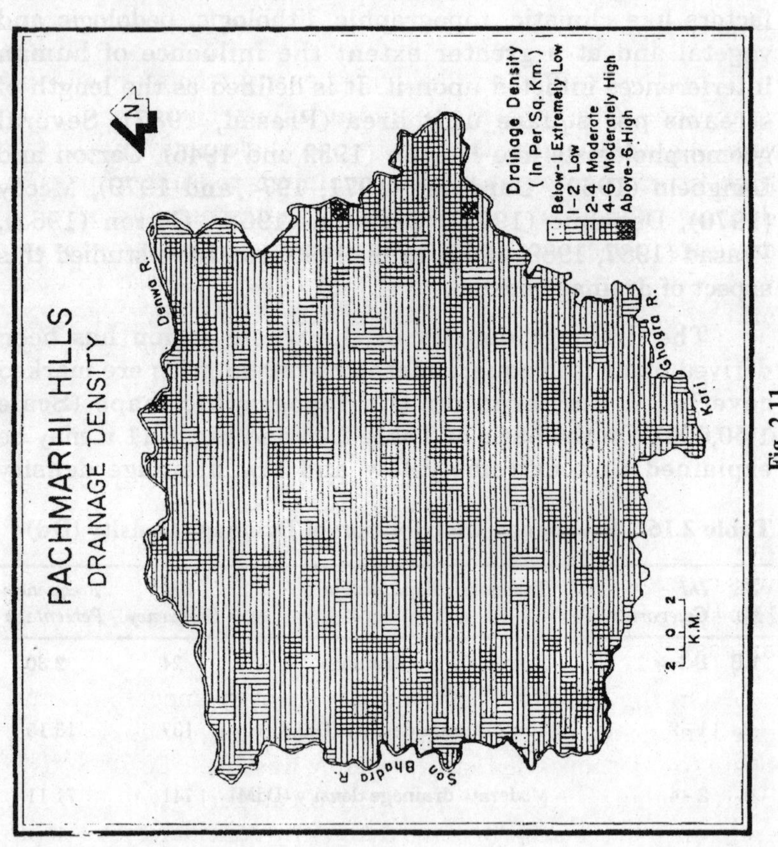
PACHMARHI HILLS
DRAINAGE DENSITY
N
Denwa R.
Ghogra R.
Kari
Son Bhadra R.
2 1 0 2
K.M.
Drainage Density
(In Per Sq. Km.)
Below-1 Extremely Low
1-2 Low
2-4 Moderate
4-6 Moderately High
Above-6 High

Fig. 2.11

Denwa and Sonbhadra river systems which form the border line from all side. Moderate drainage density registers the maximum percentage (71.11%) of grid frequencies and usually marked along the entire hill sections. Moderately high and high drainage density percentage have been noted as 13.25 per cent and 0.19 per cent respectively. The high attitudes mostly show the moderately high and high drainage density region.

On the basis of the drainage density results, it may be concluded that the high mountainous country mostly represents the higher degree of slope angles and thus shows high stream frequency, fine drainage texture and high drainage density. It is noteworthy that stream frequency, drainage density and drainage texture are the most important variables of drainage analysis and closely related to each other. Drainage texture and stream frequency are more significant parameters of dissection of terrain than drainage density.

Stream Frequency (SF)

Stream frequency or drainage frequency refers as the total number of drainage lines in a drainage basin or a morpho-unit by the corresponding area of the said unit. It was discussed by Horton (1932 and 1945) and Strahler (1957) but the proposal of the above scholars are not applicable as an easy process.

On the basis of the total number of drainage lines in a one km^2 grid area, the stream frequency of all the drainage basins and the entire study region has been derived. The derived values are classified in five categories wherein the frequency percentage are also taken into consideration (Table 2.17).

According to Table 2.17 and Fig. 2.12, it is clear that moderate group of stream frequency covers maximum percentage (66.31%) of grid frequency of the region. Very poor, poor, moderately high and high categories of stream frequency cover 2.50 per cent, 11.99 per cent, 18.43 per cent

Table 2.17 Frequency Distribution of Stream Frequency (SF)

Sl. No.	*SF Categories*	*Taxonomic Explanation*	*Grid Frequency*	*Frequency Percentage*
1.	Below 2	Very Poor Stream Frequency (SFVP)	26	2.50
2.	2 - 5	Poor Stream Frequency (SFP)	125	11.99
3.	5 - 10	Moderate Stream Frequency (SFM)	691	66.31
4.	10 - 15	Moderately High Stream Frequency (SFMH)	192	18.43
5.	Above 15	High Stream Frequency (SFH)	08	0.77
	Total		1042	100.00

and 0.77 per cent of total grid frequency of the region. From east to west the middle part of the area shows moderate to high categories of stream frequency. Very poor and poor categories together account 14.49 per cent while moderately high and high categories of stream frequency together register as 19.20 per cent of grid frequencies of the total surface area which in turn indicates that the poor and high categories of stream frequency denote more or less similar performance. The poor SF categories contain minimum areas while the minimum percentage (0.77%) is represented in high category of stream frequency.

In general, stream frequency provides more or less similar findings as observed in drainage density and drainage texture.

Drainage Texture (Dt)

Drainage texture is a fundamental geomorphic concept and an important morphometric variable of network analysis which elucidates the nature of relative spacing of drainage lines. It was discussed by Morton (1945), Smith (1950), Singh (1976), Prasad (1986, 1989 and 1990) etc. The relative spacing of the streams in an one km^2 unit area have been derived along a linear direction with the help of the following formula proposed by Prasad (1986).

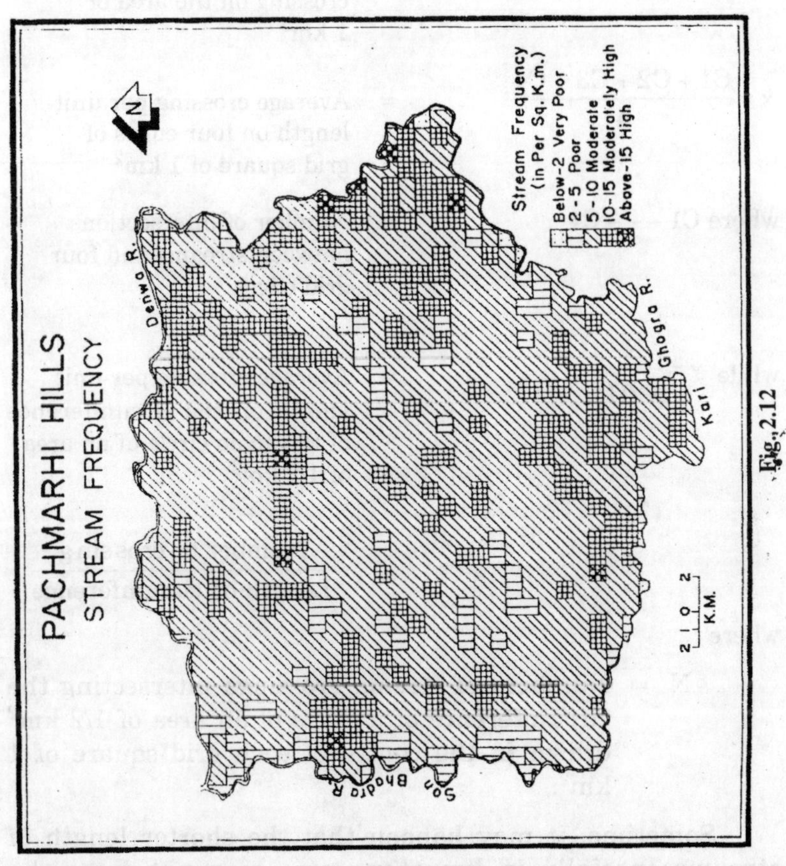

PACHMARHI HILLS
STREAM FREQUENCY
N
Stream Frequency
(in Per Sq.K.m.)
Below-2 Very Poor
2-5 Poor
5-10 Moderate
10-15 Moderately High
Above-15 High
Denwa R.
Ghogra R.
Kari
Son Bhadra R.
2 1 0 2
K.M.

Fig. 2.12

$$Dt = As = \frac{1}{(x+y)/2} = \frac{2}{x+y}$$ = Average spacing covered by one crossing ..... (1)

where Dt (As) = Drainage texture which is the average spacing of one crossing on the area of 1 km^2

$$x = \frac{C1 + C2 + C3 + C4}{4}$$ = Average crossing per unit length on four edges of grid square of 1 km^2

where C1 – – – C4 = Number of intersections between streams and four grid edges

while $y = \frac{N}{\sqrt{2\Pi}}$ = Average crossing per unit length on the circumference of the given circle of an area of 1/2 km^2

$$= \frac{\text{Number of crossing}}{\text{Length of circumference}}$$

where

N = Numbers of stream crossings intersecting the circumference of a circle of an area of 1/2 km^2 drawn in the centre of each grid square of 1 km^2.

Sometime, it may happen that the shorter length of streams specially in limestone area may not cross the circumference of a circle of an area of 1/2 km^2 drawn in the centre of grid square of 1 km^2. Therefore, it is necessary again to draw the second circle of an area occupying 1/4 km^2 within the first circle. If it is so, the above formula may be considered in the following manner (Prasad, 1987):

$$Dt = As = \frac{3}{(x + y + z)} \quad \text{... (2)}$$

where x and y are as elaborated in equation (1)

$$\text{while } z = \frac{\overline{N}}{\sqrt{\Pi}}$$

where $\overline{N}$ = Number of stream crossing intersecting the circumference of a circle of an area of 1/4 km^2 drawn within the first circle.

On the basis of the above formulae, the Dt of the region has been calculated and the frequency percentage of the derived values are given in Table 2.18. It is cartographically shown by Fig.2.13.

It is apparent from Table 2.18 that minimum grid frequency (95) and frequency percentage (9.12%) have been noted in fine drainage texture category. The highest grid frequency (503) and frequency percentage (48.27%) are obtained in moderated drainage category. Very coarse and coarse Dt categories account as 26.39 per cent and 16.22 per cent of grid frequencies of the total surface area.

Table 2.18 Frequency Distribution of Drainage Texture (Dt)

Sl. No.	*Drainage Texture Categories*	*Taxonomic Explanation*	*Grid Frequency*	*Frequency Percentage*
1.	0.2 - 0.4	Fine Drainage Texture (DtF)	95	9.12
2.	0.4 - 0.6	Moderate Drainage Texture (DtM)	503	48.27
3.	0.6 - 0.8	Coarse Drainage Texture (DtC)	169	16.22
4.	Above 0.8	Very Coarse Drainage Texture (DtVC)	275	26.39
	Total		1042	100.00

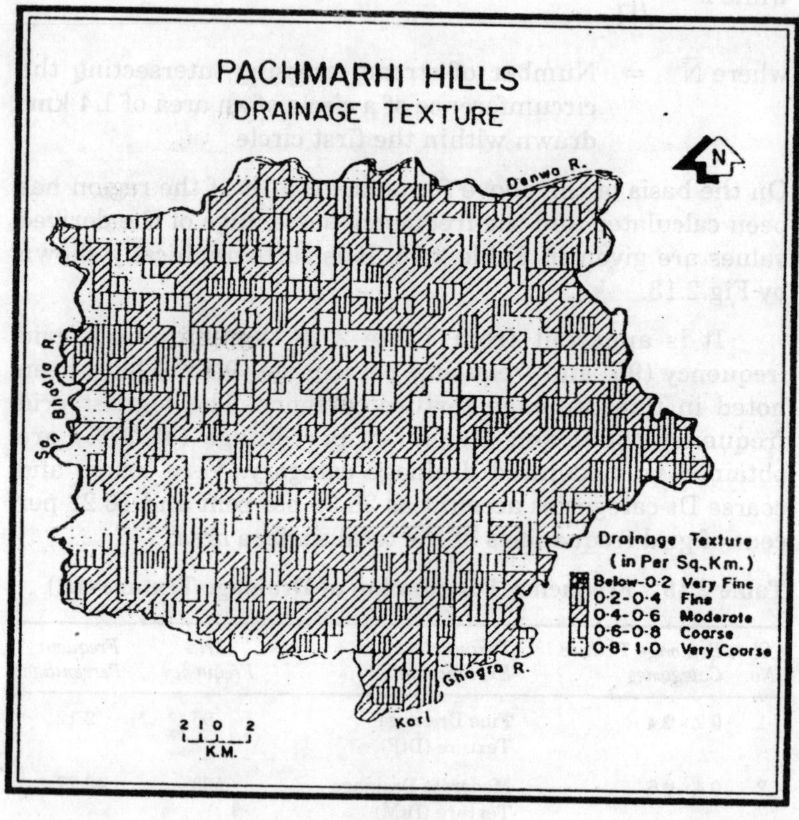
PACHMARHI HILLS
DRAINAGE TEXTURE
N
Denwa R.
Son Bhadra R.
Ghogra R.
Kori
Drainage Texture
(in Per Sq. Km.)
Below-0·2 Very Fine
0·2-0·4 Fine
0·4-0·6 Moderate
0·6-0·8 Coarse
0·8-1·0 Very Coarse
2 1 0 2
K.M.

Fig. 2.13

The choropleth map (Fig. 2.13) deals with the spatial arrangement of four drainage texture categories. On the basis of its nature and characteristics, it may be classified in the following Dt region:

(a) *Fine Dt region:* Dotted along the most of the hills of high altitudes.

(b) *Moderate Dt region:* Along the rectilinear slope profile of the region.

(c) *Coarse Dt region:* Foot hill zones where the river tributaries are comparatively less in number.

(d) *Very coarse Dt region:* Over the flat top areas and along the level valley bottom land.

AVERAGE SLOPE

The study of slope form and its evolution has become the most important aspect of geomorphology. It is an angular inclinations of terrain between hill tops and valley bottoms and as an important component of the landscape which provides not only the variety of topographical features but also makes available the evidences needed for the interpretation of the complex form of landscape (Kumar, 1979). It is directly affected by several environmental factors and also influences few of them.

The present study is based on the computation of average slope from the topographical maps on the scale 1:50,000 with 20 m contour interval through Wentworth (1930) method.

The entire region is divided in grids of one km × one km and average angles have been computed according to the following formula:

$$\text{Slope angle } (\text{Tan } \varnothing) = \frac{(\text{N}) \times \text{CI}}{0.6366 \text{ K}}$$

where

N = Number of intersections of contours per km^2 or per mile^2 length in each grid.

CI = Vertical contour interval in metre, and

K = 1000 for metric units and 5280 for feet or miles.

On the basis of the above formula and the study illustrated by Prasad (1986, 1987 and 1988), the slope angles in each grid has been obtained and classified in six categories (Table 2.19).

Table 2.19 Frequency Distribution of Average Slope (As)

Sl. No.	*Average Slope Categories*	*Taxonomic Explanation*	*Grid Frequency*	*Frequency Percentage*
1.	Below 2°	Level to very gentle slope (SL)	15	1.44
2.	2° - 5°	Gentle Slope (Sg)	61	5.85
3.	5° - 10°	Moderate slope (Sm)	269	25.82
4.	10° - 18°	Moderately Steep slope (Sms)	453	43.47
5.	18° - 30°	Steep Slope (Ss)	237	22.75
6.	30° - 45°	Very steep slope (Svs)	07	0. 67
	Table		104	100.00

On the basis of Table 2.19 and Fig. 2.14, it is clear that level to very gentle slope category occurs in very limited area (1.44%) mostly along the Denwa valley in the middle-east, northeast, middle-north flanks and along the Denwa and Sonbhadra confluence. It is also marked in the southwest of the region along the Sonbhadra river. Gentle slope is also common along the valley region and existed over 5.85 per cent of the total surface area. Moderate slope occupies the second highest grid frequency (269) containing 25.82 per cent of the total surface area. It is also very commonly marked along the marginal land of the plateau country. Moderately steep slope category registers maximum grid frequency (453) and frequency percentage (43.67%). Steep slope is marked over 22.75 per cent of the total geographical area and generally

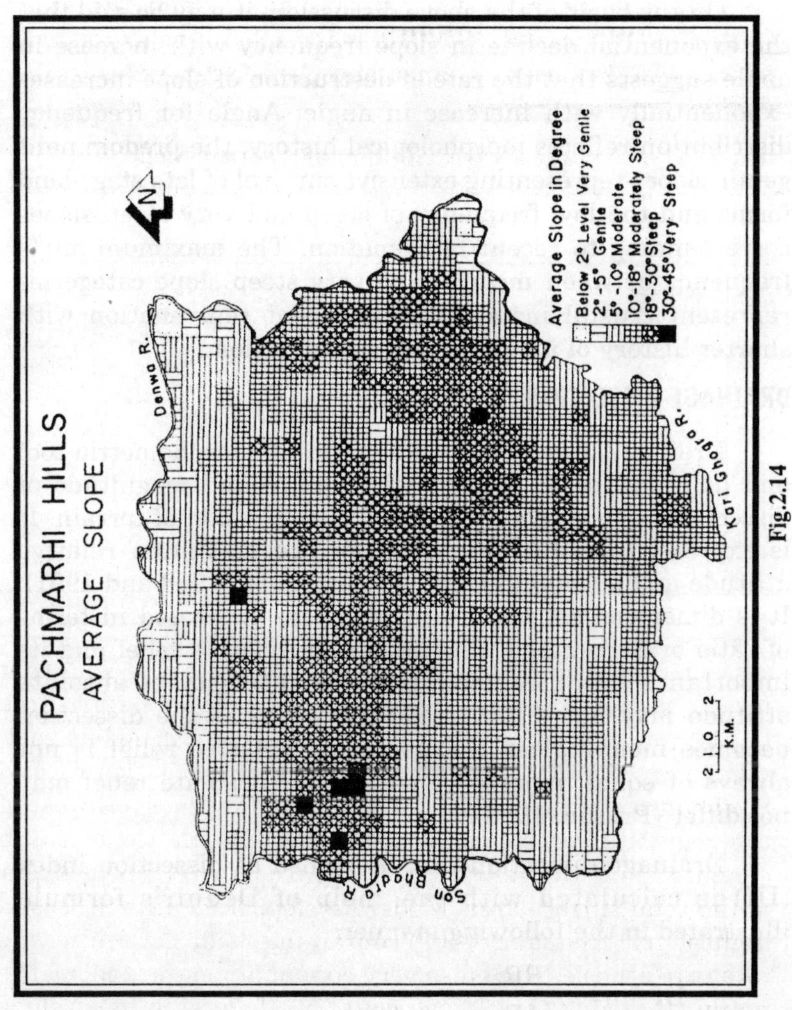
PACHMARHI HILLS
AVERAGE SLOPE
N
Denwa R.
Average Slope in Degree
Below 2° Level Very Gentle
2° -5° Gentle
5° -10° Moderate
10°-18° Moderately Steep
18°-30° Steep
30°-45° Very Steep
Kari Ghogra R.
Son Bhadra R.
2 1 0 2
K.M.

Fig.2.14

occupies the heartland of the plateau. The minimum percentage (0.67%) is marked in very steep slope category. The scattered hills top shows this type of slope angles.

On the basis of the above discussion, it may be said that the exponential decline in slope frequency with increase in angle suggests that the rate of destruction of slope increases exponentially with increase in angle. Angle for frequency distribution reflects morphological history, the predominant gentle slopes representing extensive survival of late stage land forms and the low frequency of steep and very steep slopes corresponding to recent rejuvenation. The maximum angle frequency between moderate to very steep slope categories represent crustal instability and recent rejuvenation with shorter history of its morphological evolution.

DRAINAGE DISSECTION

Drainage dissection is a significant morphometric tool and as indicator to estimate the nature and magnitude of dissection with respect to vertical exaggeration of terrain. It is expressed as the ratio between the maximum relative attitude and absolute height (Prasad, 1986, 1988 and 1991). It is dimensionless because it is always expressed in terms of ratio or percentage. The concept of relative relief and its importance becomes useless, if the intensity of absolute attitude may not differ in nature and thus the dissection becomes meaningless because equal relative relief is not always of equal importance since their absolute relief may not differ (Prasad, 1988).

Drainage dissection that is termed as dissection index (DI) is calculated with the help of Deunir's formula illustrated in the following manner:

$$DI = \frac{RR}{AR}$$

where

RR = Relative relief, and

AR = Absolute relief

The values of dissection index are basically a ratio which range between 0 and 01 but it cannot be absolute zero because no part of this planet earth is devoid of dissection except the surface which is permanently covered and protected with thick ice sheet and also it cannot be more than unity except in the case of vertical cliff.

The derived data of dissection index is classified in six categories (Table 2.20).

Table 2.20 Frequency Distribution of Dissection Index

Sl. No.	*Dissections Index*	*Taxonomic Explanation*	*Grid Frequency*	*Frequency Percentage*
1.	Below 0.1	Extremely low Dissection Index (DIEL)	170	16.31
2.	0.1 - 0.2	Low Dissection Index (DIL)	409	39.25
3.	0.2 - 0.3	Moderate Dissection Index (DIM)	302	28.98
4.	0.3 - 0.4	Moderately High Dissection Index (DIMH)	129	12.39
5.	0.4 - 0.5	High Dissection Index (DIH)	29	2.78
6.	Above 0.5	Very High Dissection (DIVH)	03	0.29
	Total		1042	100.00

Table 2.20 and Fig. 2.15 reveal the fact that extremely low, low and moderate dissection index occupy 16.31 per cent, 39.25 per cent and 28.98 per cent areas of the total surface area. Extremely low and low categories together account as 55.56 per cent area of the region. The very limited areas are registered under moderately high, high and very high dissection index categories as it is indicated by 12.39 per cent, 2.78 per cent and 0.29 per cent respectively.

CONCLUSION

Overall discussion illustrated on topo-ecology of the region shows that the region has a dense network of drainage where bifurcation ratio varies between 3.61 and 4.80. Length

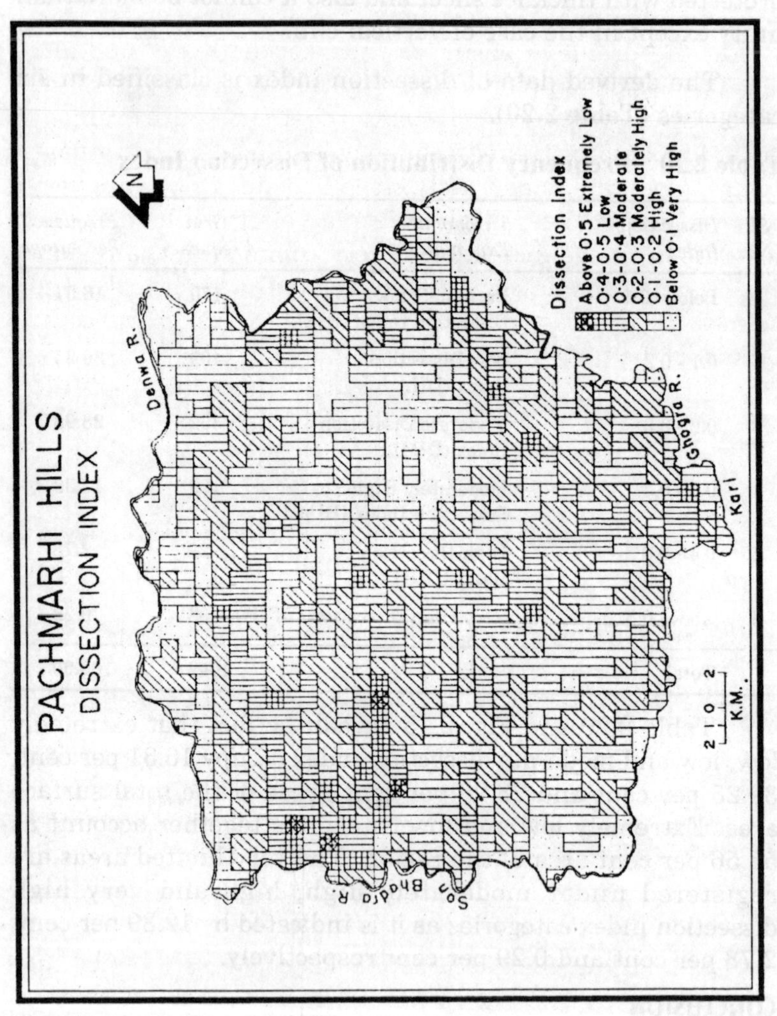
PACHMARHI HILLS
DISSECTION INDEX
N
Denwa R.
Dissection Index
Above 0·5 Extremely Low
0·4-0·5 Low
0·3-0·4 Moderate
0·2-0·3 Moderately High
0·1-0·2 High
Below 0·1 Very High
Ghogra R.
Karl
Sonbhadra R.
2 0 2
K.M.

Fig. 2.15

ratio ranges between 1.65 and 4.26 and most of the rivers show their sinuous flow path. Meandering channels are not seen due to topographic impact. Except Denwa and Sonbhadra all the rivers have their smaller areal coverage because of the tributaries of the aforesaid rivers. The area ratio of the drainage basins are marked between 4.03 and 5.54. The region is predominated by very high relative relief, moderately high and high absolute relief, moderate drainage density, moderate stream frequency, and moderate drainage texture, Moderate, moderately steep and steep slope are very common due to very high relative relief while low and moderate drainage dissection is very common in the region. Thus, the morphometric variables related to topography and drainage of the region clearly depict the real pictures as marked in the field observation.

REFERENCES

1. Cariston, C.W. 1963: *Drainage Density and Stream Flow,* U.S. Geol. Survey Professional Paper, 422C.

2. Carton, C.W. and Langbein, W.B. 1960: *Rapid Approximation of Drainage Density; Tine Intersection Method,* U.S. Geol. Surv. Wat. Res. Division, Bulletin, p.11.

3. Chorley, R.J., Malm, D.E.G. and Pogorzeiski, H.A. 1957: A New Standard for Estimating Basin Shape, *Amer. Journal of Science,* 255, pp.138-141.

4. Cotton, C.A. 1964: The control of Drainage density, *N.Z. Journal, Geology and Geophysics,* Vol.8, pp.384-392.

5. Donahue, J.J. 1972: Measuring drainage density with a dot planimeter, *Professional Geographer,* Vo1.26, pp.317-319.

6. Gardiner, V. 1971: A Drainage Density Map of Dartmoor, *The Devonshire, ASSOC,* Tr. Vol.103, pp.167-180.

7. Gardiner, V. 1974: A Photo-mechanical Techniques for Production of Drainage Density Maps, *Cartogr. Jour.* Vol. 11, pp.22-44.

8. Gardiner, V. 1979: *Estimation of Drainage Density from Topographical Variables, Water Resource Research,* No.15, pp.909-917.

9. Giusti E.V. and Schneider, W.J. 1965: *The Distribution of Branches in River Networks.* U.S. Geol. Survey, Prof. Paper 422G.
10. Horton, R.E. 1932: Drainage Basin Characteristics, *Trans. American Geophysical Union,* Vol.13, pp.350-360.
11. Horton, R.E. 1945: Erosional Development of Streams and their Drainage Basins, Hydrophysical Approach to Quantitative Morphology, *Geol. Soc. Amer. Bull.* No.56, pp.275-370.
12. Kumar, A. 1979: *Geomorphology of Simdega and its Adjoining Area,* Bihar, NGSI, Varanasi, p.50.
13. Lewin, J. 1970: *A Note on Stream Ordering, Area,* 2, pp.32-35.
14. McCoy, R.M. 1970: Automatic Measurement of Drainage Networks, *IEEE Tans. Geol. Sci. Electron,* Vo1.8, pp.178-183.
15. Melton, M.A. 1957: *An Analysis of the Relations Among Elements of Climate, Surface Properties and Geomorphology,* Office of Nawal Research, Geog. Branch, Project NR, 389-042, Technical Report II, Columbia University.
16. Miller, V.C. 1953: A Quantitative Geomorphic Study of Drainage Basin Characteristics in the Clinch Mountain Area, Va. and Tenn. Office Nawal Research Project, N.R. 389-042, Tech. Rept. 3, Columbia University.
17. Mueller, J.E. 1968: An Introduction to the Hydraulic and Topographic Sinuosity Indexes, *Ann. Ass. Amer. Geog.* 58, pp.371-385.
18. Prasad, G. 1986: Morphometric Evaluation and Interpretation of Bifurcation Ratio in Paisuni Catchment Area, Kaimur. Scarp, India, *Modelling, Simulation and Control,* AMSE Press, France, C, Vol.5, No.2, pp.1-12.
19. Prasad, G. 1986: Slope Evolution and Profile Characteristics of Chitrakut Upland, Proceeding of International AMSE Conference, Held at New Delhi.
20. Prasad, G. 1986: Morphometric Analysis of Bifurcation Ratio in Small Drainage Basin of Garara, Banda, India, *Vikassheel Bhoogol Patrika* (Hindi) Vo1.5, No.l.
21. Prasad, G. 1986: Morphogenesis of Guptgodavari Caves in Vindhyan Dolomite Rocks Near Chitrakut, M.P., India. *Modelling, Simulation and Control,* AMSE Press France, C, Vol.5, No.l, pp.31-42.

22. Prasad, G. 1986: Environmental Modelling and Morphometric Analysis of Chitrakut Upland, India, Ph.D. Thesis (unpublished) Submitted to AMSE University, France.

23. Prasad, G. 1987: Law and Order of Hydrogeological Concept Regarding the Drainage Density and Drainage Density Ratio under the Effective Environmental Control of Paisuni-Ganta Interfluvial Zones, India, *Modelling, Simulation and Control,* AMSE Press, France, C, Vol.7, No.2, pp.1-13.

24. Prasad, G. 1987: Law of Stream Number of Selected Drainage Models of Chitrakut Upland, India, *Modelling, Simulation and Control,* AMSE. Press, France, C, Vol.7, No.4, pp.25-31.

25. Prasad, G. 1987: Channel Controls: A Comparative Study of Chitrakut Sinuosity and Environmental Upland and Garara Catchment Area, Banda, U.P., India, *Modelling, Simulation and Control,* AMSE Press, France, C, Vol.7, No.4, pp.40-47.

26. Prasad, G. 1987: Measurement of Complexity of Relationship between Average Slope and Associated Morphometric Variables of Chitrakut Upland, India, *Modelling, Simulation and Control,* AMSE Press, France, C, Vol.8, No.2, pp.35-36.

27. Prasad, G. 1987: Modelling and Simulation of Relief Ratio and Environmental Controls of Morphometric Approach for 20 Drainage Models of Chitrakut Upland, India, *Modelling, Simulation and Control,* AMSE Press, France C, Vo1.8, No.3, pp.11-16.

28. Prasad, G. 1987: Geometry of Shape of Closed Links: A Morphometric Approach for 20 Selected Model of Drainage Basins of Chitrakut Upland, India. *Modelling, Simulation and Control,* AMSE Press, France , C, Vo1.8, No.4, pp.49-59.

29. Prasad, G. 1988: A Comparative Study of Relative Relief of Chitrakut Upland, AMSE Transactions, AMSE Press, France, Vo1.2, No.3, pp.1-26.

30. Prasad, G. 1988: *Measurement of Drainage Dissection and Environmental Control,* AMSE Press, France, Vol.2, No.4, pp.1-22.

31. Prasad, G. 1988: Slope, Geomorphology and Environment, *AMSE Transactions,* AMSE Press, France, Vo1.2, No.4, pp.23-30.

32. Prasad, G. 1989: Some Recent Techniques and Concepts in Geomorphological Research, The Brahmavarta, *Geographical Journal of India,* Vo1.1, No.1, pp.25-30.

33. Prasad, G. 1990: Measurement of Hortanian Law of Stream Length, *Modelling, Simulation and Control,* AMSE Press, France, C, Vol.23, No.3, pp.11-22.

34. Prasad, G. 1991: Law of Basin Area, Area Ratio and Allowmetric Growth, *Modelling, Simulation and Control,* AMSE Press, France, C, Vol.23, No.3, pp.13-22.

35. Prasad, G. 1991: Morphological Processes, Land Forms and their significance in Environmental Management of Chitrakut and Adjoining Region, Ph.D. Thesis submitted to Kanpur University, Kanpur.

36. Prasad, G. 1993: Measurement Techniques and Concepts of Drainage Density, Drainage Density Ratio and Microrelief density, *H.B.B. Patrika,* Vol.29, No.l-2, pp.17-23.

37. Prasad, G. 1995: Measurement Techniques and Some Recent Concepts of Drainage Density Ratio (B), *The Indian Geographical Journal,* Madras, Vol.70, No.2, pp.122-123.

38. Schidegger, A.E. 1965: The Algebra of Stream-order Numbers, U.S. Geological Survey, Prof. Paper 525B, B1, pp.87-89.

39. Schidegger, A.E. 1966: Stochastic Branching Processes and the Law of Stream Orders, *Water Resources Research,* 2, pp.199-203.

40. Schumm, S.A. 1956: The Evolution of Drainage Systems and Slopes in Badlands at Perth Amboy, *New Jersey, Geol. Soc. Amer. Bull.* 67, pp.597-646.

41. Schumm, S.A. 1963: Sinuosity of Alluvial Rivers on the Great Plains, *Geol. Soc. Amer. Bull.* 74, pp.1089-1100.

42. Singh, S. 1976: On the Quantitative Parameters for the Computation of Drainage Density, Texture and Frequency: A Case Study of a Part of the Ranchi Plateau, *National Geographer,* Vol.10, No.1, pp.21-31.

43. Singh, V.P, 1984: A Morphometric Study of Terrain of Patlands of the Chhotanagpur Region, Ph.D. Thesis (unpublished) Awadh University, Faizabad, pp.184-210.

44. Shreve, R.L. 1966: Statistical Law of Stream Numbers. *Journal of Geology,* 74, pp.17-37.

45. Shreve, R.L. 1967: Infinite Topologically Random Channel Networks, *Journal of Geology,* 75, pp.178-186.

46. Smart, J.S. 1967: A Comment on Horton's Law of Stream Numbers, *Water Resource Research,* 3, pp.773-776.

47. Smith, K.G. 1950: Standard for Grading Texture of Erosional Topography, *American Journal of Science,* 248, pp.566-568.

48. Strahler, A.N. 1957: Quantitative Analysis of Watershed Geomorphology, *Ader. Geophy. Union,* Tr.38, pp.913-920.

49. Strahler, A.N. 1964: Quantitative Geomorphology of Drainage Basins and Channel Networks, Appeared in *Handbook of Applied Hydrology* Edited by Chow, V.T. pp.4-39.

50. Strahler, A.N. 1969: *Physical Geography,* John Wiley and Sons, Inc. p.483.

51. Wentworth, C.K. 1930: A Simplified Method of Determining the Average Slope of Land Surfaces, *American Journal of Science,* 20, pp.184-194.

Morphogenetic Processes

MEASUREMENT OF INTENSITY OF MORPHOLOGICAL PRÓCESSES

The study area generally experiences dry climate under which different landscapes have been formed due to dynamic cycle of fluvial processes associated with mechanical and chemical weathering. Due to lack of advanced instrumental facilities and some other mathematical techniques to measure the nature and magnitude of morphological processes, the intensity of processes, operating on the terrain of the region, has been determined with the help of suggested models of Peltier (1950) and Wilson (1973) depending upon the mean annual temperature (21.76ºC) and mean annual rainfall (1596.0 mm) recorded in 1996 of the region.

Interpretation of Peltier's Model

According to Peltier's model seven graphs showing the relationships between various geomorphic processes and climatic parameters of mean annual precipitation and temperature have been drawn and location of the study region is marked with a dot (Fig. 3.1). Thus, the various types of weathering (Fig. 3.1A), the intensity zone of chemical weathering (Fig. 3.1B), mechanical weathering (Fig. 3.1C), Mass movement (Fig. 3.1D), fluvial erosion (Fig. 3.1E), aeolian action (Fig. 3.1F) and the recognition of morphogenetic regions (Fig. 3.1G) incorporate the following results:

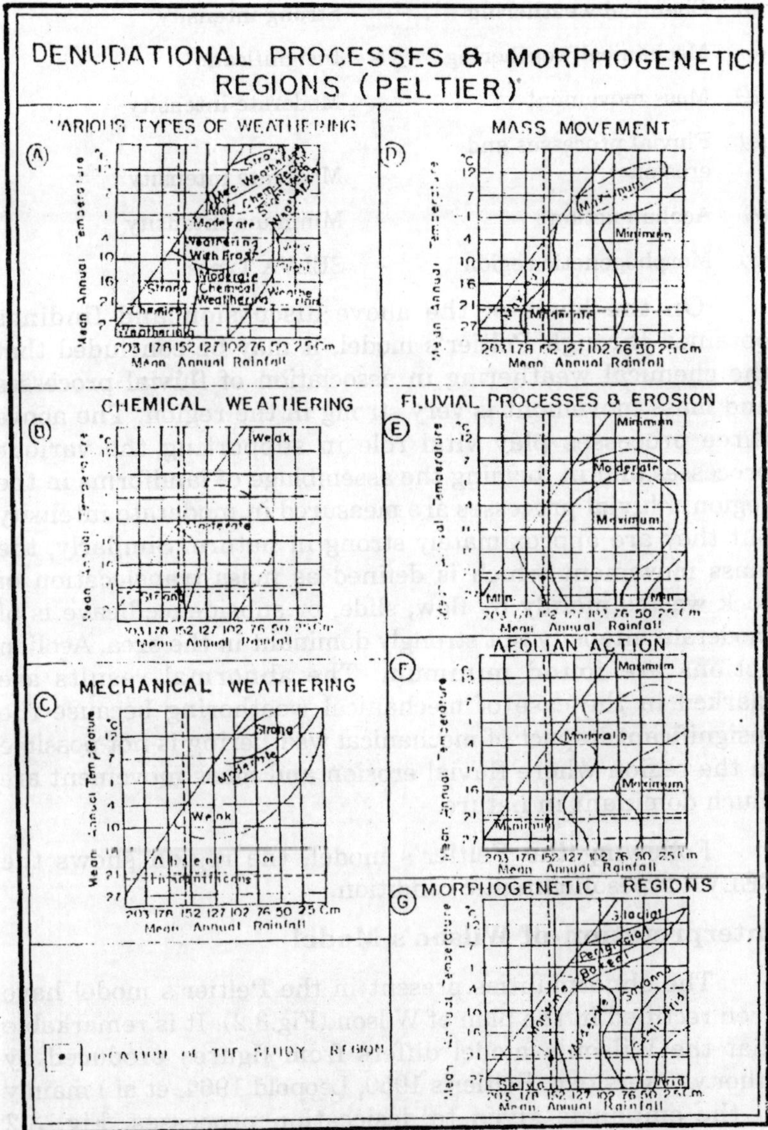
DENUDATIONAL PROCESSES & MORPHOGENETIC REGIONS (PELTIER)
VARIOUS TYPES OF WEATHERING
(A)
Mean Annual Temperature
Mean Annual Rainfall
Moderate
Weathering With Frost
Moderate Chemical Weathering
Strong Chemical Weathering
203 178 152 127 102 76 50 25 Cm
CHEMICAL WEATHERING
(B)
Weak
Moderate
Strong
MECHANICAL WEATHERING
(C)
Strong
Moderate
Weak
Absent or Insignificant
MASS MOVEMENT
(D)
Maximum
Minimum
Moderate
FLUVIAL PROCESSES & EROSION
(E)
Minimum
Moderate
Maximum
Min.
AEOLIAN ACTION
(F)
Maximum
Moderate
Minimum
MORPHOGENETIC REGIONS
(G)
Glacial
Periglacial
Boreal
Maritime
Moderate
Savanna
Semi-Arid
Selva
Arid
LOCATION OF THE STUDY REGION

Fig. 3.1

(a)	Types of weathering with intensity	–	Strong Chemical Weathering
(b)	Chemical weathering	–	Strong intensity
(c)	Mechanical weathering	–	Insignificant
(d)	Mass movement	–	Moderate intensity
(e)	Fluvial processes and erosion	–	Moderate intensity
(f)	Aeolian action	–	Minimum intensity
(g)	Morphogenetic region	–	SELVA Types.

On the basis of the above discussion and findings obtained through Peltier's model, it may be concluded that the chemical weathering in association of fluvial processes and mass movement is very strong in the region. The above three processes play vital role in supporting the various processes and in forming the assemblage of landforms in the region. Fluvial processes are measured in moderate intensity but they are approximately strong in nature. Similarly, the mass movement which is defined as mass translocation of rock wastes mainly by flow, slide, or subsidence/heave is of moderate intensity but strongly dominant in the area. Aeolian actions are found minimum. The abnormal results are marked in the case of mechanical weathering because the insignificant impact of mechanical weathering is not possible in the region where fluvial erosion and mass movement are much dominant in nature.

Following the Peltier's model, the region shows the SELVA types of climatic condition.

Interpretation of Wilson's Model

The abnormalities present in the Peltier's model have been rectified in the plan of Wilson (Fig.3.2). It is remarkable that the Wilson's model differs from figures produced by other workers (e.g. Peltier's 1950, Leopold 1964, et al.) mainly in the emphasis given to desiccation processes. Fig. 3.2 illustrates the five determinants of morphological processes e.g. mechanical weathering (Fig. 3.2A), chemical weathering

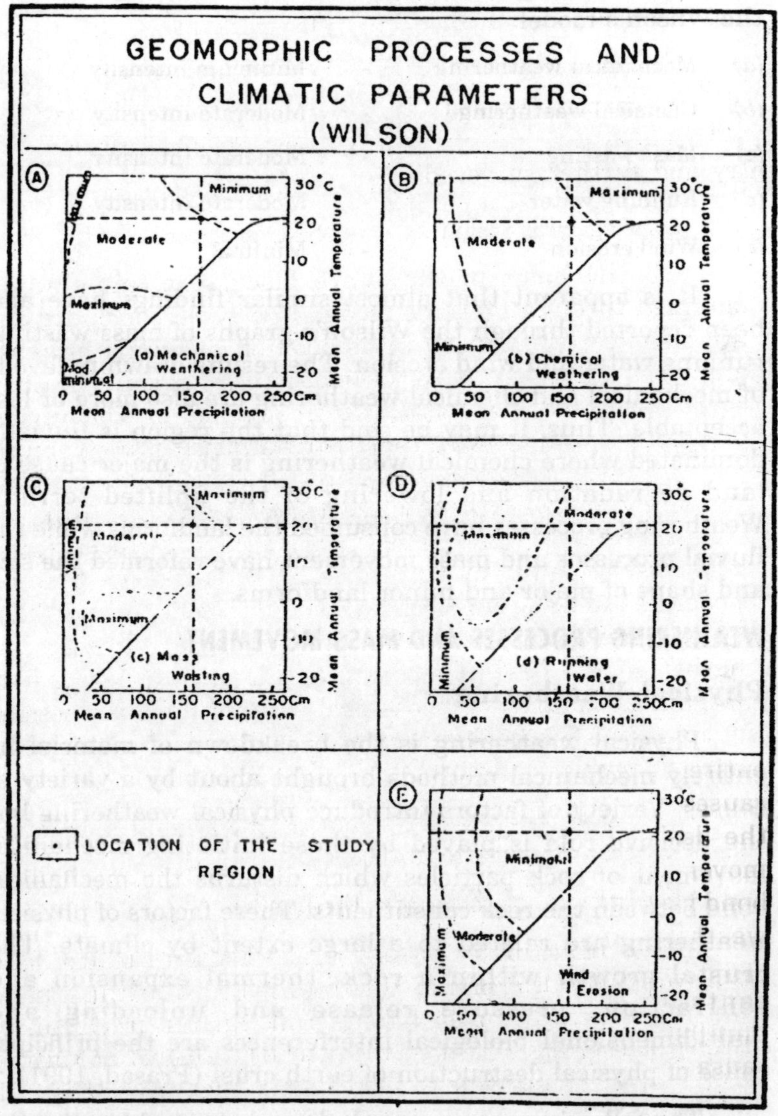
GEOMORPHIC PROCESSES AND
CLIMATIC PARAMETERS
(WILSON)
A
Minimum
Moderate
Maximum
(a) Mechanical
Weathering
Mean Annual Temperature
Mean Annual Precipitation
B
Maximum
Moderate
Minimum
(b) Chemical
Weathering
C
Maximum
Moderate
Minimum
(c) Mass
Wasting
D
Moderate
Minimum
(d) Running
Water
E
Minimal
Moderate
Maximum
Wind
Erosion
LOCATION OF THE STUDY
REGION

Fig. 3.2

(Fig. 3.2B), mass wasting (Fig. 3.2C), running water (Fig. 3.2D) and wind erosion (Fig. 3.2E) respectively.

The following results have been drawn on the basis of the Wilson's model:

(a)	Mechanical weathering	-	Minimum intensity
(b)	Chemical weathering	-	Moderate intensity
(c)	Mass wasting	-	Moderate intensity
(d)	Running water	-	Moderate intensity
(e)	Wind erosion	-	Minimal

It is apparent that almost similar findings have also been reported through the Wilson's graphs of mass wasting, running water and wind erosion. The results drawn in favour of mechanical and chemical weathering are also more or less acceptable. Thus, it may be said that the region is fluvially dominated where chemical weathering is the major cause of land degradation and lowering of the uplifted terrain. Weathering processes have consumed the landforms while the fluvial processes and mass movement have reformed the size and shape of major and minor landforms.

WEATHERING PROCESSES AND MASS MOVEMENT

Physical Weathering

Physical weathering is the breakdown of material by entirely mechanical methods brought about by a variety of causes. Variety of factors introduce physical weathering but the decisive role is played by those initiating mechanical movement of rock particles which disturbs the mechanical bond between the rock constituents. These factors of physical weathering are related to a large extent by climate. The crustal growth within a rock, thermal expansion and contraction, pressure release and unloading and multidimensional biological interferences are the principal cause of physical destruction of earth crust (Prasad, 1991).

The following processes relating to physical weathering are recognised during the extensive field study of the region.

(a) Sheeting

Sheeting is the division of rock into sheets or beds by joint like fractures generally found parallel to the ground surface. This process is also termed as the name of 'Topographic Jointing'. Sheeting appears during the period of uplift and erosion when the expansion of rock masses suddenly develop the cracks. The sheeting is first recognised by Gilbert (1904) in granite country and further surveyed by Bradley (1963) in massive sandstone, Ollier 6 Tuddemham (1962) in massive arkose and conglomerate, Kiersch and Asce (1964) in limestone and by Currecy (1968) in bedded sandstone. Sheeting is also noticed in glaciated regions by Chapman and Rioux (1958), Lewis (1954) etc.

In the present study region, sheeting is noticed over the massive sandstone of Pachmarhi and Deccan trap lava deposits. The sheeting in the sandstone country is caused by overburden pressure released during the early period of tectonic movement. This type of examples can be easily traced near Ganesh hills in the north of Pachmarhi urban area where massive sandstone rocks are horizontally bedded. Vertical sheeting and disintegration of rocks beds can be observed along the free face element of Jata Shanker Pahari. Hummock sheeting is seen along the sandstone country in the way of Rajat fall. Similarly well developed sheeting of sandstone rocks are found near Apsara Vihar while lava sheeting is seen over the sandstone rocks near Apsara Vihar. The rolling surface of Deccan lava deposits is also the best example of sheeting in the eastern flank of Pandav cave hills.

(b) Unloading

There are minute differences between sheeting and unloading. Unloading causes sheeting. In the case of unloading, horizontal and sub-horizontal partings of rocks can be recognised generally towards the top of the exposures. It is notable that the unloading also results vertical cracks but it is found less dominating. The sheeting are more or less due to unloading process. Intersecting unloading planes may be observed at some places. These fractures suddenly appeared after the release of confining pressure. Unloading is also progressed by erosional agents. The area of Jambudeep river, near Jata Shanker hills, the rocky terrain in the way of Rajat falls and located on Bharna Pahar, the opening of Ganesh hills before Jata Shanker hills and the collapse of limestone rocks along the foothill zones of Jambudeep Pahar clearly indicate the examples of unloading.

(c) Spalling

Spalling is the process which produces spalls of platy rock fragments generally Lozerge shaped or irregular. They break from the wall by a combination of tension cracks parallel to the unloaded surfaces and shear fractures due to compression acting parallel to the wall. Following to Ollier (1969) spalling exists due to physical conditions favour cracking and by spatial considerations which prevent the formation of sheets. In spalling, the unloading produces stress differences which in turn cause uniaxial, biaxial and radial bending stresses. The field measurement shows that each of the bending stress easily can be found. Here it is notable that sheeting, unloading and spalling processes jointly act at one place. These are interdependent to each other.

(d) Block Separation or Disintegration

Block separation causes due to physical properties of rocks. In the sedimentary rocks weathering enlarge the rock joints and so more of the rock surface exposes to the atmosphere suddenly breaking up the rock under block separation process. This type of weathering can be recognised over sandstone, limestone and granite structure of the region.

(e) Spheroidal Weathering

Some well joined rocks decompose readily the corners and edges are attacked more than the flat surfaces. Thus the rocks becomes rounded in all three dimensions. As each shell is weathered away, a fresh shell of rock is exposed and the rock decomposes becoming more sphere like as weathering progressess. This type of weathering process is called as spheroidal weathering. In this condition, the rocks show globular jointing. The top surface exposures of Jambudeep Pahar shows this type of weathering.

(f) Columnar weathering

Columnar jointing in granite and sandstone rocks initiates columnar weathering. Rocks contract and jointing are generally found in vertical manner. The weathering process attacks the corners and edges of the rocks most probably in the upper portions.

(g) Exploitation

Temperature changes cause expansion or shirnkage of rocks. In the case when heating agent is sunshine, the said process is termed as insolation weathering. The heating and cooling generally depend on the nature of the rocks. The continued process of temperature change initiate unequal expansion and contraction of the rocks or peel away the surface layers from the parent rocks. This phenomena is known as exploitation or destruction and desquamation of rocks. Such type of weathering is measured near Jata Shanker along the river Jambudeep and also downward sites of Jata Shanker mandir.

(h) Slaking and Splitting

Alternate wetting and drying of rocks can be a very important factor to weathering, a process known as slaking. The two types of disintegration of rocks may be possible, a minor disintegration and a major disintegration of rocks. The minor is caused due to slaking and the major is initiated by splitting in two or more large pieces of about equal size. Cracking generally intended to concentrate along the bedding planes and cleavage planes when these are formed. This type of weathering processes can be observed in fine graine rocks.

(i) Mechanical Collapse

Rock weathering initiates the form of mechanical collapse following undercutting of various sorts. Most of the steep scarps denoting gorge positions mark the undercutting of hard rocks and thus mechanical collapse is very common there. Mechanical collapses are marked along Jambudeep basin and in the lower reaches of Bee fall. This type of processing is most common in the entire region. Mechanical collapse fortifies the mass movement.

CHEMICAL WEATHERING

Chemical weathering has a deeper penetrative capacity and its impact is defined by a radical transformation of rocks. Chemical weathering consists the effect of various atmospheric factors; like oxygen, carbondioxide, air, moisture and the active organic substances produced by the vital activity of plant and animal organisms (Prasad, 1987). The major reactions causing chemical weathering are as oxidation, hydration, dissolution and hydrolysis. The measurement of chemical reaction is not possible but boldly can be traced in the entire region because the study area is thickly vegetated and the running water activities are most common. Most of the spring conditions, sites of water falls, river sides and dolomite caves generally indicate chemical weathering.

The larger area of the region experiences eluviation or leaching which in turn initiates rain wash or humification process. Creep or solifluction are the other important microprocess. The PH value is measured more than 7 which indicates that the sandstone is a calcite dominated and largely affected by solution actions. View of Gupta Mahadeo, located at Mahadeo hills, indicate the solution of dolomite rocks and sites of larger caves of Jata Shanker in dolomite rocks indicate solution action and mechanical collapse of rocks. Several solution pits and cavities near Apsara Vihar along the river Bainganga are also marked while the foothill zones of Pandav cave hills marked the solution cavities. The ravination process which is most common in the region have caused the greater areas as ravinous land with the help of running water bodies.

BIOTIC WEATHERING

Plants, animals and bacteria largely control the breakdown of rocks and minerals. Polynov (1937) believes that completely sterile weathering is impossible. Vegetation litter and decaying vegetation—various stages of humus are important in helping to conserve moisture which in turn enhance weathering. Another important effect of vegetation is the formation of 'leaf leachates' which are very important in cheluviation. The tree leaves actively mobilise iron than grasses. The vegetation controls the erosion and its important role is to ease the removal of weathered products. According to Walker (1963) the erosion is at least twice fast under grass and it is most likely that weathering therefore also be faster.

The dominant role of vegetation on weathering of rocks is very much important; physically and chemically. Small plants like grasses, bushes and thorny scrubs which are most common over the isolated hills, ravine zones and on the plateau country are more or less moisture loving. The rocks upon which they took their birth has kept continually moist. The moisture or acidic humus acts in various ways in decomposing and disintegrating the rocks. Most of the tree grains entrance into small cracks and crevices of the rocks continue to grow in chinks. Growing roots exert strong pressure on the walls of cracks and acting as wedges pry them apart into slabs and fragments. The dead roots of plants become swollen with rain water and also widen the cracks. They give rise to certain organic compounds during decomposition which attack the associated rocks and dissolve some of their constituents.

Few examples of biotic weathering can be quoated from the different part of the region. The sites in the way of Rajat fall, Jambudeep Valley, Bharna Pahar, Apsara Vihar, Pandav caves, Steep gorges near Handi Khoh, Ganesh hills, the way of Bee fall and Ganesh hills show the notable examples of biotic weathering.

The impact of human interferences, the effects of animals, worms and insects on weathering is also most common throughout the region.

MASS MOVEMENT AND ASSOCIATED PROCESSES

The term mass movement is applied to those processes which involve a transfer of slope forming materials from higher to lower ground under the influence of gravity without the primary assistance of a fluid transporting agent. They merge imperceptibly with processes in which transporting media, such as air water or ice are involved and which are generally called mass transport processes. Thc movements may be slow or rapid, shallow or deep and include one or more of the mechanisms of creep, flow, slide or fall (Brunsden, 1979).

Mass movement occurs when the disturbing forces become greater than the resistance of the slope forming materials. Mass movement can be based upon the three modes of movement namely flow slide or subsidence/heave. Hutchinson (1968) has distinguished four types of mass movement phenomena namely creep, frozen ground phenomena, landslides and subsidence. Landslides, rock falls, slumps and surface wash are the processes actively engaged in the region. The most of the hilly terrain associating cliffing cause mechanical collapse and mass movement.

FLUVIAL PROCESSES

The recognition of fluvial processes and its intensity have been considered as a difficult task for geomorphologists therefore the study relates around the realm of secondary data, topographical outlooks and general field survey.

The overall comments on erosional typology and techniques have been postulated on the basis of the field observations concluded from time to time. The review of system and mechanism of fluvial processes incorporating infiltration, length of overland flow and surface flow have been made with the help of data collected for 12 drainage basins. The interpretation of surface flow is stated through the measurement of sinuosity indices of 12 drainage basins. The remaining processes like pipe flow, subsurface flow, sheet wash and sheet flow and rill wash and gullying have been discussed on the basis of the field survey.

The geological work of following surface water confirms the fluvial processes and fluvio-culture of the region. Its intensity relates to mass of water and flow velocity. As mentioned earlier, Pachmarhi and its adjoining areas are the part and parcel of Satpura range which existed from Satpura basin. The processes which are operating today are not the same as operated in the whole span of time of Satpura history. They differ in nature due to variability in climatic conditions. The geologists approach on stratigraphy of Pachmarhi plateau formed in different geological time also confirm the changing operational characteristics of the processes. But the up-to-date Satpura formation clearly suggests the dominance of fluvial processes through the passage of time. Following the scenary of present day landforms and climatic conditions, the entire region is considered as fluvially dominated terrain.

The intensity of present day processes operating the existing terrain have been determined with the help of Peltier's (1950) and Wilson's (1973) models (Figs. 3.1 and 3.2) depending upon the mean annual temperature and mean annual rainfall of the region. In this measurement, the region shows the climatic conditions of SELVA types where fluvial processes and chemical weathering are marked at maximum intensity. Thus, the present day landforms are shaping themselves under the effect of fluvial processes. The stratigraphic sequence of Pachmarhi Plateau has gone in a changing marine conditions by active palaeo-currents with a significant swift in climatic regime. Since the existence of the mountain, fluvial processes accelerate their geological works and a thick amount of alluvium deposited in Denwa valley and its major tributaries. Although the alluvium cover of Denwa and its tributaries are nominal than that of Narmada valley, yet the alluvium layer of Denwa denotes the operational history of running water. It is notable that Satpura formation feels dominance of marine processes in its early stage and experiencing several breaks of climatic change and geological formation but after mountain building in the region, it registers the active operation of fluvial processes. Present day erosional activity differs from basin to basin due to mass and velocity of water and from one litho-structure to another due to nature of composition of rocks. Thus, sheeting, piping, rilling, gullying etc. processes exist at different dimensions following the riverine society.

Geologists approach impressed the views that the geological history of the region have experienced climatic crux in phases. The marine environment as agitated by turbulent palaeo-current was thought to be warmer, humid, arid and semi-arid in nature where regression, transgression, exhumation, denudation, winnowing, abrasion, shifting of land and sea and emergence and submergence of sea level have technically progressed the suits of stratigraphical sequence with the march of geo-chronology. These techniques imprisoned in basin structural history undoubtedly at the time of mountain building of Satpura ranges. Since existence, the uplifted land mass has undergone several tectonic jerks and still stands has faced the impact of exogenetic processes. The denudational processes and several other erosional techniques cycled the plateau more or less in senile stage. The transporting agencies have made more complex terrain profile plan controlling with erosional, transportational and depositional techniques. The present day assemblage of Landforms of Pachmarhi recite the above fact.

The records of geologists observations highlight the formational and deformational techniques of mountains. Ray and Bhattacharya (1982) opined that due to gradual uplift and change in palaeogeomorphic setup the present day topography has evolved. The uplifted topography was later subjected to process of denudation, scarp retreat and peneplanation giving rise to the landforms. Quite later, the Satpura formations and topography were covered under profuse outpourings of the Deccan lavas. Denudational processes have exhumed the Satpura topography which continues back-wearing of scarps even to the present day. Several times, it was uplifted and peneplained in several phases. The evidence of peneplanation is afforded by the presence of many flat topped hills at different levels with the same elevation. It is observed during the field survey that the scarpfaces occur vertical cliffing generally underlying rectilinear or concave slope profile in plan. The free face element of the scarp is actively backweared by parallel retreat processes. The erosional processes as suggested by Davis, i.e. slope declining is also progressed here because most of hills have diminished or lowered at a greater extent. The rectilinear slopes of the scarps which are systematically covered with erosional sediments are found in ruined form due to rilling and ravaging impact. Thus, the erosional capacity of fluvial processes are variable in nature from one place to another.

The scientific observation of system and mechanism of fluvial processes may be discussed in the following manner.

Infiltration

Infiltration may be simply defined as the process of water entering the soil. It is not similar to hydraulic conductivity of soil which affects the rate of infiltration but it may be expressed as the flux of water across the soil surface (Prasad, 1991). Robin (1966) has identified the rain infiltration on the basis of water into soil differs with time. Clowes and Comfort (1982) believes that the rate can be quite high at the onset of rain because of the dry soil. The small voids between clay and silt particles reduce the infiltration rate. In comparison sand and gravel have high infiltration rates as a result of the large voids between the particles. Thus the infiltration capacity is influenced by a number of factors like rainfall intensity, duration, drop size and soil characteristics (texture, structure, depth nature and proportion of clay minerals, vegetation and landuse).

The region receives 1810.57 mm average annual rainfall wherein degree of intensity of infiltration index is also noted variable from place to place. The infiltration index ranges between 15 per cent and 25 per cent. The hard rock structure marked on the plateau which have small voids between the two particles reduce the infiltration rate. On the contrary the sedimented loose sand, gravel clay and morrum in Denwa valley having large voids between the particles have allowed high rate of infiltration. The several cave positions where dolomite rocks is most common, the infiltration of rain water is also measured high.

Water transmitted between peds (soil structure) as well as particles. It is observed in plateau region that a well developed grump structure has numerous well connected voids which have allowed the water movement in all directions. In the case of blocky structures having regular small size voids, the water moves in all directions. Prismatic structures poses large peds with well defined voids having a dominant downward orientation and water movement in vertical directions. Platy structures are seen having generally large peds with ill-defined voids possessing a dominant lateral water movement. The development of crusting, surface erosion, water repellency, accessions of salt in the profile at the end of summer season and specially freezing of water in the sub-surface, all cause wide temporal variations in the rate of infiltration. Vegetation covers also causes a considerable amount of seasonal variability. The presence of vegetation tends actually to encourage infiltration by inducing high soil water deficits.

Overland Flow

Overland flow is that part of the lateral inflow which flows overland surface towards a stream channel. The length of overland flow is one of the most important independent variables affecting both the hydrologic and physiographic development of drainage basins (Fig. 3.3, Horton, 1945). It is sustained by a relatively thin layer of surface detention. It can be used to describe the length of flow of water over the ground surface before it becomes concentrated in definite stream channels or permanent drainage channels. Length of

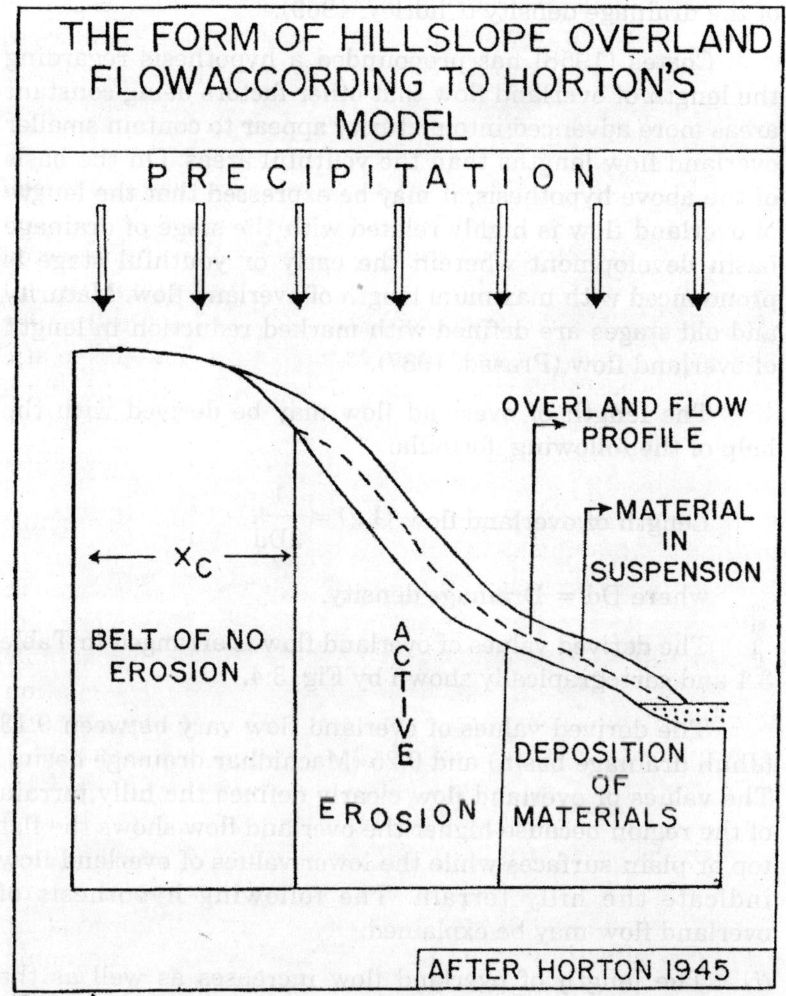
THE FORM OF HILLSLOPE OVERLAND FLOWACCORDING TO HORTON'S MODEL
PRECIPITATION
OVERLAND FLOW PROFILE
MATERIAL IN SUSPENSION
X_C
BELT OF NO EROSION
ACTIVE EROSION
DEPOSITION OF MATERIALS
AFTER HORTON 1945

Fig. 3.3

overland is considered as the mean horizontal length of flowpath from the water divide to the stream in the first order basin and is a measure of stream spacing the degree of dissection. It is approximately one-half of the reciprocal of the drainage density (Chorley, 1969).

Coates (1958) has propounded a hypothesis regarding the length of overland flow that other factors being constant areas more advanced into maturity appear to contain smaller overland flow lengths than the youthful areas. On the basis of the above hypothesis, it may be expressed that the length of overland flow is highly related with the stage of drainage basin development wherein the early or youthful stage is pronounced with maximum length of overland flow. Maturity and old stages are defined with marked reduction in length of overland flow (Prasad, 1987).

The length of overland flow may be derived with the help of the following formula:

$$\text{Length of overland flow } (\bar{L}g) = \frac{1}{2Dd}$$

where Dd = Drainage density.

The derived values of overland flow is arranged in Table 3.1 and cartograpically shown by Fig. 3.4.

The derived values of overland flow vary between 0.13 (Jhuli drainage basin) and 0.25 (Machidhar drainage basin). The values of overland flow clearly defines the hilly terrain of the region because higher the overland flow shows the flat top or plain surfaces while the lower values of overland flow indicate the hilly terrain. The following hypothesis of overland flow may be explained:

(i) The length of overland flow increases as well as the channel and ground slopes decrease over the homogeneous lithological controls of the region.

(ii) Lithological heterogeneity and steeper the channel and ground slopes cause lower the values of overland flow

Table 3.1 Length of Overland Flow ($\bar{L}g$)

Sl. No.	*Drainage Basins*	*Average Drainage Density (Dd)*	*Length of overland flow ($\bar{L}g$) in Kms.*
1.	Kumajhiri	3.21	0.16
2.	Nishan	2.99	0.17
3.	Kabra	2.91	0.17
4.	Machidhar	2.04	0.25
5.	Ganjakunwar	3.25	0.15
6.	Sawania	3.50	0.14
7.	Jhuli	3.90	0.13
8.	Nagdwari	2.90	0.17
9.	Jamna	3.11	0.16
10.	Bori	2.91	0.17
11.	Denwa (link)	2.86	0.17
12.	Bainganga	3.10	0.16

and much drainage dissection mostly over high hilly terrain.

(iii) Higher the values of length of overland flow older the stage of basin development while lower the values of $\bar{L}g$ indicate early stage of terrain development.

The above hypotheses are applicable in 12 drainage basins of the region. All the basins took their birth upon high altitude and also drain over the high mountainous country and so the values of $\bar{L}g$ as marked are very low.

Concentrated through Flow or Pipe Flow

Pipe flow occurs generally by those soils which have either peaty surface horizons and impermeable layers at shallow depth or those where surface horizons are loamy and the slope is steep. Piping has been recorded in many areas of the world. It provides a macropore networks for the quick transmission of through flow. Jones (1971) has marked several interesting areas of piping in the British isles. Gilman and Newson (1980) have also investigated few examples of pipe networks on upland wales.

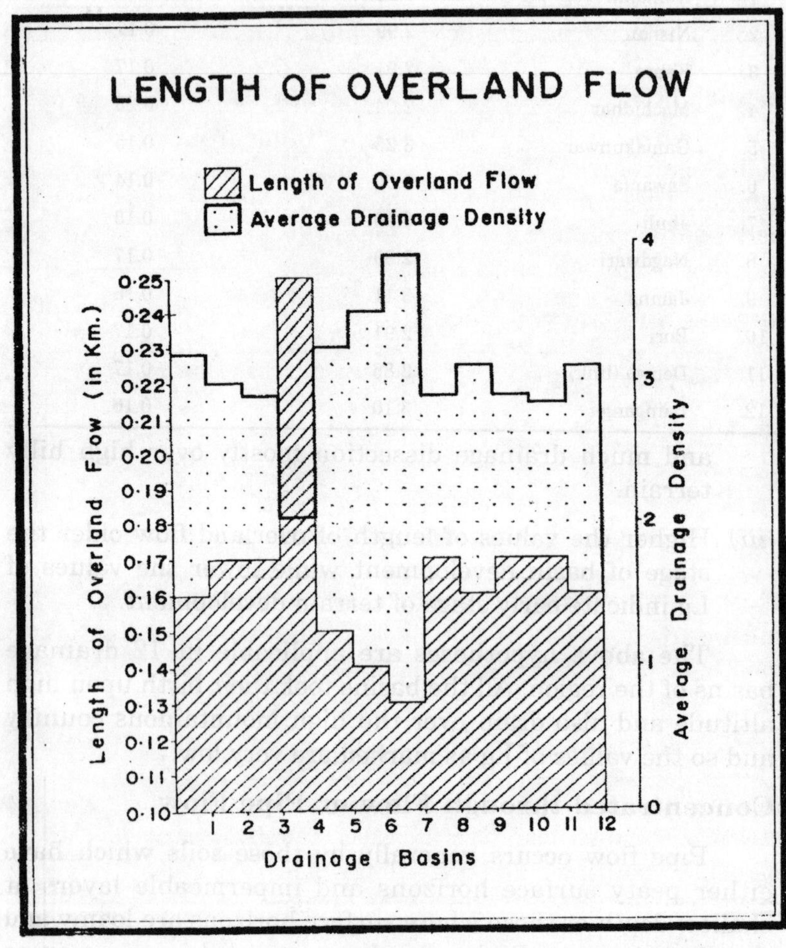
LENGTH OF OVERLAND FLOW
Length of Overland Flow
Average Drainage Density
Length of Overland Flow (in Km.)
0·25
0·24
0·23
0·22
0·21
0·20
0·19
0·18
0·17
0·16
0·15
0·14
0·13
0·12
0·11
0·10
1 2 3 4 5 6 7 8 9 10 11 12
Drainage Basins
Average Drainage Density
4
3
2
1
0

Fig. 3.4

Eluviation and dessication are the most important processes which produce pipes. Eluviation allows the fine fraction to be removed progressively through the soil matrix. In the present study, the pipes are measured along the river sides and over the rectilinear slope profiles throughout the region. The dessication cracks are most common in black soil areas. The steep valley exposing gorge position near Handikhoh, the rectilinear slopes covered with green plants near Chaura Darshan known as Malcompoint, Valley slope profile near Malcompoint steep valleys and rectilinear slopes near Chaura Darshan and the deep soil profile along the link road approaching Little fall generally show the peculiar examples of concentrated through flow.

Surface and Sub-surface Flow

Water moves over the ground surface across slope in variety of ways. The processes like rain wash, surface wash, solution and mass movement exist by flowing water. Wind action is also considered as surface processes. The relative importance of these processes and their interactions are largely but not entirely governed by climate. Schumm (1956 a and b) has suggested that the interaction of the various processes in any specific site can be extremely complicated. Unconcentrated flow in thin sheets is only possible on fairly smooth surfaces. Emmett (1970) found that sheet flow or wash varied greatly in character and depth and was a mixture of laminar and turbulent flow. Clowes and Comfort (1982) express that the flowing water particles along an open channel does not follow a direct straight line or even a smoothly curving path or trajectory. It moves vertically and laterally in addition to its overall down stream direction. The actual velocity of this particle is recognised much greater than the mean velocity in a down stream direction. This type of flow is called as turbulent flow and predominates in most natural river channels. A second type flow is laminar flow exists by the linear trajectories of water particles. It is generally found at lower flow velocities and is of very limited occurrence in natural river channels. It is usually found that the fastest flow of water occurs in the centre of the channel. The edges of the channel show slowest surface flow. The formation of rills along hill sides and river beds divert the surface flow into many channels. This type of examples can be seen along the rectilinear slopes of the major hills of the region. Deepest valley of Bainganga river supported by rectilinear slopes of Bharna Pahar and consumed by rills, Giddli river valley supported by isolated hills of Marodeo Pahar, well dissected by rills showing the declining slope profile of Handikhoh Pahar and a view of Nishangarh, Brijlaldeo and Nandigarh Pahar pictured from Dhupgarh Pahar and drained by Michijhiri Jambudeep and Lingi drainage system generally show the major hills where rectilinear slope profiles are well dissected by several rills from top to bottom. Due to rilling and gullying processes the slope forms have been mirrored by multidirectional surface flow. The site known as Denwa Darshan where Denwa the most prominent river of the areas, flows between the steep gorge position of Pachmarhi hills, is beautiful. The surface flow of Jambudeep river near Ganesh hills, a opening of Jambudeep Pahar near Pachmarhi town is also remarkable. The flowing water also receives the largest amount of city dirty water by supporting channels which in turn cause pollution in the associated areas. It is also remarkable that the surface flow is well governed with seasonal rainfall variations. Mostly the small tributaries of the region have no water during the summer season except Denwa, Sonbhadra and Bainganga. The surface flow of the tributary streams where generally water fall exists is totally governed by spring conditions and seepages existed in the hilly terrain of the region.

The topographic variations, lithological changes and climatic uncertainty of the region have caused so many surface flow paths (fingertip tributaries) along the slope dynamics. The drainage network (Fig. 1.6) and drainage orientation (Fig. 1.7) of the region truely explain the surface flow complexity and orientation. It is also well recognised that the upland region denotes quick run-off for a short period of time while the lower sections of the valleys indicate maximum quantity of flowing water generally for a calendar year but it denotes lesser numbers of surface tributaries.

In order to measure the surface flow regularity and the variability raised due to different environmental factors, it has been attempted to find out the morphometric results of various sinuosity indices of 12 drainage basins of the region (Table 2.7). According to Table 2.7, it is clear that there are not a single river which has its meandering course because the sinuosity index of all the basins range between 1.09 and 1.3 which is the indicative of sinuous pattern of the rivers. The hydraulic dominance of the rivers have been also noticed through morphometric observations. Only Machidhar, Sawania, Nagdwari and Denwa link rivers exhibit the highest percentage of topographic sinuosity index and thus advocate the dominance of topography in their catchment areas.

It is observed in the field study that the surface flow paths are generally straight in nature over the hard and compact geological structure of Gondwana Sandstone.

On the basis of the above analysis, it may be stated that surface flow velocity and directions are not regular in nature. Sub-surface flow is found dominant in soluble dolomite rocks. Various caves as noticed along the scarp zone validate the above fact. Bada Mahadeo, Nagdwari, Denwa Darshan, Bariam Pahar and so many caves of the region are of the indicative of sub-surface flow in the existing geological structures of dolomite. The another examples of sub-surface flow processes like rilling, piping, tunnelling and gullying can be easily seen in the ravinous belt of the region.

Sheet Wash and Sheet Flow

Sheet wash and sheet flow is the processes initiated by accelerated soil charged water to continuous surface flow of water swills the land surface and causes the finer soil particles to be washed away. This swilling is called sheet wash (Rabinson, 1972). Continued sheet wash will lead to the removal of a thin layer of soil which is known as sheet erosion.

A continuous thin layer of overland flow is called as sheet flow. It washes mainly the particles of fine silt and clay down the slope. In the present study, it is found that the sheet wash is dominating in the lower sections of the slope, rolling and smooth surfaces of the area along the valley sides and most of the parts of concavity formed along the isolated hills. The affective surface wash is measured under the catchment areas of Denwa and Sonbhadra rivers.

Rill and Gullying

Rill wash and gullying is a dominant fluvial processes. It is found in survey that the rectilinear slopes of Pachmarhi hills are badly wounded by rilling and gullying processes. The initiation of rills can be traced in most of the localities where rills are in normal conditions.

CONCLUSION

The assemblage of landforms and the terrain of Pachmarhi plateau is governed by a complex suite of weathering and fluvial processes. The relative importance of these processes is largely a function of climatic conditions. Ravination appears to be the most dominating processes along the hill sides and river courses. Various forms of weathering are noted in moderate and strong conditions. Surface and sub-surface processes are also dominant over the weaker sections of rocks forming the upland country. Sub-surface solution is peculiar in dolomite rocks and as well as sandstone structures. Mass movement is prominent features due to steep valley sides and gorge positions in the area. These factors are related directly or indirectly to topographic influences. Slope gradient is probably a major controlling factor and on most slopes the processes vary systematically with position. Thus, climate, topography slope and lithological characteristics appear to be the dominant controlling factors of the region.

REFERENCES

1. Bradley, W.C. 1963: Large Scale Exfoliation in Massive Sandstone of the Colorado Plateau, *Bulletin of Geological Society of America*, 74, pp.519-528.

2. Brunsden, D. 1979: *Mass Movement Appeared in Process in Geomorphology* (ed. by Embleton and Thornes, J.), Arnold Heine Mann Publishers (India) Pvt. Ltd., p.130.

3. Chapman C.A. and Rioux, R.L. 1958: Statistical Study of Topography, Sheeting and Jointing in Granite, Acadia National Park, Maine, *American Journal of Science,* 2 pp.111-127.

4. Chorley, R.J. (ed.), 1969: *Water, Earth and Man,* Methuen, London.

5 Clowes, A. and Comfort, P. 1982: *Process and Landform, Conceptual Frameworks in Geography,* Oliver and Boyd, Edinburgh.

6. Currecy, D.T. 1968: Bell Field Dam, Victoria Pt. 1 Site Geology, Inst. Engineers, Aust. Ann. Conf. Papers, pp.133-136.

7. Emmett, W.M. 1970: The Hydraulics of Overland Flow, U.S. Geol. Survey, *Professional Papers,* 662, p.46.

8. Gilvert, G.K. 1904: Domes and Domed Structures of the High Sierra, *Bulletin of Geological Society of America,* 15, pp.29-36.

9. Gilman, K. and Newson, M.D. 1980: Soil Pipes and Pipe Flow, British Geomorphological Research, Group Research Monograph, 1, Norwich, Geo-Abstracts.

10. Horton, R.E. 1945: Erosional Development of Streams and their Drainage Basins, Hydrophysical Approach to Quantitative Morphology, *Bulletin of Geological Society of America,* 56, pp.275-370.

11. Hutchinson, J.N. 1968: *Mass Movement in Fairbridge, R.W. The Encyclopaedia of Geomorphology,* Reinhold, New York, pp.688-696.

12. Jones, J.A.A. 1971: Soil Piping and Stream Channel Infiltration, *Water Resource Research,* 7, pp.602-610.

13. Kiersche, G.A. and Asce, F. 1964: Vaiont Reservoir Disaster, Civil Engineering, 34, pp.32-39.

14. Lewis, W.V. 1954: Pressure Release and Glacial Erosion, *Journal of Glaciology,* 2, pp.417-422 .

15. Ollier, C. 1969: Weathering, English Language Book Society and Longman Group Limited, London, p.1.

16. Ollier, C.D. and Tuddemham, W.G. 1962: Inselbergs of Central Australia, Zeit. F. Geomorph. 5, pp.257-276.

17. Peltier, L.C. 1950: The geographical cycle in periglacial regions as it is to climatic geomorphology, *Annals,* Association of American Geographers, 40, pp.214-236.

18. Polynov, B.B. 1937: *The Cycle of Weathering,* Murby London, Trans. Alexander, Muir, p.230.

19. Prasad, G. 1987: Impact of Length of Overland Flow on the Hydrologic and Physiographic Development of Drainage Basins 'Under Major Environmental Controls of Chitrakut Upland, India, *Modelling, Simulation and Control,* C, Vol.7, No.3, AMSE Press, France, pp.30-37.

20. Prasad, G. 1987: Channel Sinuosity and Environmental Controls: A Comparative Study of Chitrakut Upland and Garara Catchment Area, Banda, U.P., India, *Modelling, Simulation and Control,* C, Vo1.7, No. 4, AMSE Press, France, pp.40-47.

21. Prasad, G. 1987: Modelling of Morphogenetic Processes of Chitrakut Upland, *AMSE Transactions,* Vol.1, No.2, AMSE Press, France, pp.1-12.

22. Prasad, G. 1987: Measurement of Natural Hazard by Ravination and Land Degradation Processes on Chitrakut Upland, India, AMSE Transactions, Vol.1, No.3, AMSE Press, France, pp.49-63.

23. Prasad, G. 1991: Introduction of Fluvio-erosional Ecology, AMSE Transactions, Vol.9, No. 2, AMSE Press, France, pp.1-10.

24. Prasad, G. 1991: Morphological Processes, Landforms and their Significance in Environmental Management of Chitrakut and its Adjoining Region, Ph.D. Thesis (unpublished) submitted to Kanpur University, Kanpur.

25. Robinson, H. 1972: *Biogeography,* Macdonald and Evens, London.

26. Roy, A.K., Bhattacharya, A. 1982: Regional Geomorphology of Vindhyachal, Appeared in *Geology of Vindhyachal,* ed. by Valdiya, Bhatia and Gaur, Hindustan Publishing Corpn., New Delhi, pp. 9-22.

27. Schumm, S.A. 1956: The Role of Creep and Rain Wash in the Retreat of Badlands Slopes, *American Journal of Science,* 254, pp. 693-706.

28. Walker, E.H., 1963: Relative Rates of Erosion Under Grass and Forest in a Valley of Western Wyoming Northwest, *Science*, 37, pp.104-111.

29. Wilson, L. 1973: Relationships Between Geomorphic Processes and Modern Climates as a Method in Palaeoclimatology, Appeared in Climatic Geomorphology (ed. Derbyshire, E.) *Geographical Readings*, Macmillan Press, London, pp.269-278.

Morphogenetic Landforms

In the fundamental concepts of geomorphology, it is well postulated that the geomorphic processes operating at differential rates leave their distinctive imprint upon landforms. It is also a conclusive remarks of geomorphic concepts that each geomorphic process developed its own characteristics assemblage of landforms (Prasad, 1987) with a full appreciation of the manifold influences of the geologic and climatic changes during the passage of time. If it is geomorphic truth that the different erosional agents act upon the earth surface, produce an orderly sequence of landforms, it may be undoubtedly said that weathering processes including fluvial processes produce various suit of landforms. Thus the nature of weathering and erosional activities and its associated landforms are now the fundamental question of present discussion.

WEATHERING AND THE EVOLUTION OF LANDFORMS

Landforms due to Constant Volume Weathering

As mineral alteration results in the formation of new minerals of lower density than the originals, it is usually assumed that the mineral expands on weathering. Rocks being composed of minerals are assumed to be expand likewise when weathered. Commonly when a rock of the earth's surface it does expand, the change in volume causing distinctive features. However, a great deal of weathering occurs with volume change. Constant volume weathering takes place at some depth below the earth's surface (Ollier, 1969).

Ollier believes that the structures of the original rocks found perfectly preserved despite extensive chemical and mineral alteration in most of the areas of extensive deep weathering. He reported some examples of small quartz veins where joints, small faults, original bedding and other structural features are seen in original form in the weathered materials. Thus, in this context, it is possible to find out the evolution of landforms without volume change on the basis of the field measurement of the region and with the help of the Ollier's experiments, the following landforms produced due to constant volume weathering can be recognised.

(a) Constant Regolith and Saprolite

It is most common by deep weathering results that rock may be weathered to great depths. Rotten rock in place can be termed saprolite. The regolith covers both residual and transported loose or soft material overlying solid bed rock. It is observed that the profile contains several zones below the ground surface but the materials of regolith and saprolite have not changed their volume but weathered minutely due to pressure, temperature and side impact. This type of example can be seen near the side exposures of rocks near Pandav caves hills and the rocks exposure along the footpath of Beefall. These exposures illustrate residual debris of structureless sand clay, rounded, angular, prismatic and interlocked corestones with loose morrum deposits. Small cracking without gap, colouring and decomposition in upper mineral grains can be minutely observed here.

(b) Spheroidally Weathered Blocks

Spheroidal weathering generally initiates expansion of rocks. The spheroidally weathered blocks can be exposed by chemical migration of elements within the rocks. Hydrolysis in the chief weathering agent of spheroidal shape of rocks structure. Spheroidal shape occurs due to most rapid alteration at edges and corners of blocks which in turn with advancing exfoliation causes the original angular joint block to a rounder and spheroidal shape.

Spheroidal weathering landforms appear in the following conditions:

(i) Small boulders at the surface may be weathered from all sides by flaking process and simulate spheroidal weathering. In this case the unconfined boulders would probably expand. Sandstone boulders existed in the Jambudeep valley near Jata Shanker hills clearly show the spheroidal shape of the boulders at ground surface. It is also reported over the hills top of Bharna and Patarkota Pahars and boldly exposed near Denwa Darshan

(ii) In the second case spheroidally weathered boulders may be surveyed at the same depth below the ground surface. Due to subsequent erosion, the spheroidal rock surface may be exhumed at the ground surface. Ollier suggests that these boulders do not constitute proof of formation of spheroidally weathered boulders at the surface. Once exposed at the surface, their course of weathering is likely to change to a volume increase alteration.

These types of spheroidally weathered boulders may be observed along most of the residual hills of the upland. These types of spheroidally weathered boulders may be traced in Jambudeep valley near Ganesh hills and Jata Shanker hills and near the Denwa Darshan along the left flank of Pachmarhi-Pipariya road.

(c) Colour Banding

Colour banding appears in porous rocks (sandstone, limestones and mudstones) due to alternating enrichment and depletion mostly of iron oxides. Periodic; precipitation mostly affects these landforms. Ollier has observed that the banding may result from the patchy and periodic drying up of groundwater with precipitation at the surface of diffuse water bodies.

The colour' bands may be formed in different shape and size either following the edges of joint blocks or irregularly in various directions. They may be like chit and chips, circular, elongated, prismatic shape etc. The bands vary in thickness according to nature of porosity of rocks and the intensity and impact of periodic precipitation.

The examples of colour banding are well visualized in the limestone structure of Jata Shanker cave and Gupta and Bada Mahadeo caves. The investigator denotes that these are formed due to continuous groundwater precipitation along the edges of the dolomite blocks. The bands are generally marked along the edges of the rocks in a form of chit and chips. This condition is most common in Bada Mahadeo cave.

(d) Joint Hardening and Softening

Joint hardening and softening depend upon the nature of the rocks and its joints and fissures characteristics. Water moves with dissolved ions along the joints and fissures of the rocks which in turn initiate weathering processes. If the impregnated joints are harder than the blocks, later erosion may give rise to box work type of feature. This type of joint hardening and softening is visualized in the sandstone rocks near Jata Shanker hills. It is also well marked at Apsara Vihar. Jambudeep Pahar also exhibits the joint hardening and softening of rocks.

EXPANSION WEATHERING AND CAUSATIVE LANDFORMS

Expansion of rocks causes due to temperature changes. Thermal expansion and contraction of rocks cause disintegration. Dark minerals absorb heat faster which may also give rise to differential expansion leading to many small stresses within a rock. Differential expansion may lead to the formation of minute cracks and possibly granular disintegration. The landforms of expansion weathering take form in the following conditions.

Landforms due to Unloading

Unloading is the process of massive exfoliation (dilatation) that gives rise to slabs of rock several feet in thickness and associated landforms (Ollier, 1969). These landforms are mostly recognised in granites and partially seen in other igneous rocks, metamorphic and in some sedimentary rocks. Unloading may cause the following landforms.

(a) Unloading Sheets

Unloading process initiates to produce a large and thick curved slabs of rocks which contains an exposure of the dome-shaped rock beneath. These sheets may be termed as beds or joints. It is also known as topographic jointing. These partings are mostly found according to sedimentation and parallel to the ground surface.

The plateau region illustrates several examples of unloading sheets. The sites near Jata Shanker and Ganesh hills clearly denote the unloading sheets of sandstone rocks.

(b) Unloading Domes

Unloading domes is the name given to those convex hills of hard rock capping which are usually devoid of vegetation and partly covered with broken unloaded sheets. The Deccan lava deposits of Pandav caves hill shows unloading domes.

DIFFERENTIAL WEATHERING AND ASSOCIATED LANDFORMS

Features due to differential weathering are numerous. Differential weathering occurs by structural or lithological differences in rocks. Irregular weathering processes also initiate landform differentiations on homogeneous rock characteristics. Random patches of vegetal properties on different lithological units may also give rise the assemblage of landforms due to differential weathering. Thus, it may be said that differential weathering can occurs on all sides. At the upper end of the scale differential weathering and erosion make birth of valleys and hills of multiple character. It registers various micro-level regional landforms which are not very easy to recognise in the field measurement. It is also remarkable that both micro and macro weathering landforms

have their equal significance. Micro features may be several in number on one macro landforms. Thus the small features are the cause and effect of the larger sized landforms. Several geomorphologists have suggested some example of micro features like ground level platform (near Pandav cave), boulder streams, weathering pits, box wark, honey combed structure, hoodeo, butcher block, raised rim, ball and buns, gobular mass of differential weathering etc. produced by differential weathering. The measurement of micro features is difficult task and not presented in the text.

Landforms due to Sculping

(a) Sculpted Granite Blocks

Sculping only refers to the surface differential weathering but not subsurface alteration. Many corestones become sculpted when they are exposed to surface weathering and erosion. It causes an individual boulders into a variety of shapes that almost defy classification. The shape, spacing, orientation of depressions and rises seem to be of endless variety (Ollier, 1969). It occurs generally in granite rocks and the intensity of weathering appears quite different in the same area. A peculiar example of sculping in harder sandstones and Deccan lava deposits can be seen in the larger blocks existing near Jata Shanker hills and over Bharna Pahar.

LANDFORMS DUE TO STRIPPING

Due to stripping, blocks of unweathered rock may float in the regolith as corestones. The following landforms are thought to be formed by stripping processes.

Tors and Related Landforms

Following to Ollier (1969), Tors are small hills or peaks of boulders usually about 20 to 60 feet high rising abruptly from the surrounding gentle ground surface. Tors appears during the course of erosion when the basal surface of weathering may be exhumed and the corestoness left behind as the soft material of the regolith is washed away. Stripping and subaerial weathering cause the formation of tors and associated landforms. It may be formed by one cycle but many are formed by two cycle processes of deep weathering followed by exhumation. Tors are most common as an isolated exposure of much joined hard rock, standing as a prominent castellated mass above the general surface of a plateau (Prasad, 1991).

Tors may be formed in different climate at different height. The size and shape of the boulders may vary according to rock structures and tors forming processes.

Tors, one of the most controversial landforms, are piles of broken and exposed masses of hard rocks particularly granites having a crown of rock blocks of different size on the top and clitters (Trains of blocks) on the sides (Singh, 1977). Characteristically, tors are groups of spheroidally weathered rock boulders rounded in bed rock, which may be exposed as a basal rock surface or concealed by a waste mantle. Although sometimes confused with stacks, the rounded shapes and the way in which the blocks are piled up or perched in rather improbable positions on the massive unweathered rock base (Faniran, 1971a) distinguish tors form these other landforms.

Tors are generally seen in the area over Deccan lava deposits. They are most common in Gondwana sandstones. Rounded tors are visible over Bharna and Patarkota Pahars. Near Denwa darshan Tors are found in different size and shape. Cylindrical tors are observed near Jata Shanker hills. Domical or Millstone tors can be marked in the way of Rajat fall, in Jambudeep valley, over Ganesh hill while cuboidal and elongated tors are seen in the lower reaches of Jambudeep valley. Tors may be existed on the ground surface, along the rectilinear slopes, on the hill tops and along the valley sides. The aforesaid examples validate the fact clearly.

Stripping Sinkholes and Caves

Sinkholes are the depression generally caused by

underground water solution over the dolomite rocks. But stripping gives rise some peculiar sinkholes which in turn cause cavity along the joints. Sinkholes are very common over dolomitic structures of Jata Shanker hills and Mahadeo Parvat. A peculiar example of sandstone cave is seen in Pandav cave hills. It is well developed in Gondwana sandstone occupied by Deccan lava deposits.

LIMESTONE WEATHERING AND SOLUTIONAL LANDFORMS

Some parts of the Pachmarhi plateau shows dolomitic structures underneath the Gondwana sandstone which causes solutional landforms of the following types.

Terra Rossa

Terra rossa is a term applied loosely to any fine textured and fairly plastic clay that acquires a natural vitreous skin in burning and that is used in the manufacture of terra-cotta. It contains a peculiar brownish-red or yellowish-red colour. The descending groundwater usually leaves a residue of a red clayey soil mantling the surface and extending down into open joints in limestone area by surface and near surface solution. The loose materials are known as Terra-rossa. It may be found on the rock surface of moderate and gentle slope in varying thickness. Gupta Mahadeo Pahari indicates yellowish-red colour of terra rossa resting surrounding the limestone rocks,

Cave

Cave formation is a typical topography in limestone area. The formation of cave depends upon the solution of groundwater and thereby caused weathering impact of existing limestone. The nature and characteristics of limestone rocks are also the dominant factors of cave formation (Prasad, 1986). In the present study region, caves are the prominent features in dolomitic structure observed near Jata Shanker hills, Bada Mahadeo, Gupta Mahadeo, Bariam Pahar and Nagdwari hills (Fig.4.1). Another caves of the region are known as Khoh and discussed under fluvial landforms.

Cave Platform

Cave platform may be existed in hard and compact dolomite rocks due to solution of groundwater during the downward movement of water level. Solution is seen prominent where weaker formation existed along the harder rocks and thus the harder mass of dolomite rock, may be existed as platform surrounded by eroded dolomite surfaces due to downward movement of groundwater in the cave formation. This type of platforms may be seen in Jata Shanker cave and also in the inner part of the Bada Mahadeo cave.

Stalactites and Stalagmites

Stalactites are the pendant mass of calcite hanging vertically from the roof of a cave. It is a cylindrical or conical deposit of mineral matter sedimented from drops of water containing calcium bicarbonate which have seeped through crevices and joints of limestone rocks. The roof of the Bada Mahadeo cave generally forms several stalactites. It is also visible in Jata Shanker cave.

Stalagmites are the similar mass to a stalactites growing upward from the floor of cave and below the stalactites by driping water. It is also composed of calcite. In all the caves of Pachmarhi only loose materials of calcite are noted in agglomeration in place of stalagmites on cave floor.

It is remarkable that the caves of Jata Shanker hills and Gupt Mahadeo indicate well developed 'DRIP STONE' along the cave roof in spite of stalactites. The cave roof and walls are also marked with columns. 'Flow stone' and 'Rim stone' are common features where stone moves with flowing water in different size and shape.

Helictite

Davis (1930) suggested another feature in limestone cave namely as helictite which is most common to stalactite. Helictite may be in formed by deposition of calcite from solutions percolating down fine fissures onto ceilings of cave usually where a strong current of air is present. It may be thin as a thread, in spirals or loops or festooned with tenacles. It is believed that the growth of helictite does not necessarily extend along vertical lines. Its individual parts may grow upward, horizontally, obliquely or like a curves as well as downward. It is not certain why helictites are apparently able to defy gravity but the most logical explanation seems to be that they develop where water is not entering the cave in sufficient quantity to give rise to drops that fall. If there is only enough water to keep the surface wet, then growth of the individual parts of the helictite depends upon change orientation of the crystal axes of calcium carbonate. This may be in any direction.

The present investigator believes that most of the stalactites as discussed earlier are helictites because very few of them contain water. The hanging calcites of caves are most probably helictite.

FLUVIAL TOPOGRAPHY

Fluvial Processes and Development of Slopes and Scarps

The study region is fluvially dominated where fluvial processes have actively consumed the land and surface irregularities have initiated at a greater extent. There are 4296 streams of Ist order, 946 streams of IInd order, 185 streams of IIIrd order, 44 streams of IVth order, 9 streams of Vth order and only one stream of VIth order respectively. The total surface area of 908 km^2 registers 4.73 drainage lines at per km^2 which is also the indicative of the dominance of fluvial processes.

The present day topography as represented by ridges, plateaus and scarps are the results of upliftment, structural deformation and denudation through ages. The present day slopes and scarps have also generated through parallel retreat and backwearing of the initial landscape. The residual hills of lower height group and their slopes are the result of slope declining processes. The scarps and the detached hills generally show that flat top surfaces underlying vertical cliffing of limited extent. This situation indicates the parallel retreat of the free face elements and in a long duration backwearing of the scarp faces (Prasad, 1986). Most of the scarps show high degree of slope angles along their free face elements. The lower section of the profiles are usually found rectilinear in plan because the transporting agencies have deposited loose materials along the slopes. Convexo-concave plans are also visible where the vertical cliffing of dissected hills are diminished due to slope declining processes. Thus, it may be said that the slopes of the region have been caused due to parallel retreat and backwearing processes at higher height area and slope declining processes have extensively consumed the landscape greatly in detached hills areas. It is also remarkable that these slopes are caused due to erosional activities but the low degree of slope angles along the lower sections of the plateau rim have been initiated due to depositional processes. The lowering of the slope is also caused by sediment accumulation in valley bottom. These slopes are principally governed with existing drainage of the region. Overall discussions reveal the facts that the slopes whether erosional or depositional are formed by fluvial processes of the region.

The development of the scarps are principally governed with slopes initiated by erosional processes. The plateau rim is highly dissected. The degree of dissection is evident from the choropleth map of dissection index (Fig.2.15). The sites of Ganesh hills, Bee fall, Hill side slope of Chaura Darshan, Chaura Pahar, Pandav caves, Rajat fall area, Dhupgarh Pahar, and Jata Shanker hills generally show the nature of scarps forming various elements of slope due to denudational processes. The various slope profile like convexo-freeface rectilinear, vertical freeface convexo rectilinear concave, convexo-rectilinear, convexo-concave etc. are visible in the complex hill country of the region.

The field survey reveals the fact that the steep scarps prove the parallel retreat of slopes with sufficient loss in gradient. Even the flat rolling surfaces between the detached hills and the plateau rims may not be taken as a result of lateral planation rather they are the outcome of back wearing of hill sides due to parallel retreat of slopes.

The most outstanding feature of the scarps is that they are under effect of continuous process of parallel retreat. The recession of the scarps has taken place and is still active under the impact of weathering of the scarpfaces and constant removal of debris. Various detached hills exhibiting the example of Mesa, Butte etc. projecting above the general rolling uplands are the left over remnants of the recession of the scarps.

Thus, the slope evolution of the present day topography can be explained by composite theory of parallel retreat, slope replacement and slope declining. Single process cannot be formed thc complexity of slopes of the region.

Other Erosional Landforms

(a) Mesas and Buttes

Mesa is a flat-topped eminence (Table land) existing as a remnant of denudation of a plateau and commonly capped with a resistant rock stratum. Butte is a small isolated portion of Mesa. These Landforms exist between the assemblage of detached dissected hills. Most of the area away from the scarplines register several peculiar examples of detached hills and define the nature of Mesas and Buttes.

A panoramic view of Ganesh hills and Chauragarh hill show the several examples of Mesas and Buttes.

(b) Water Divides

Water divides are the source regions (birth place) of tributaries flowing according to slope. It is determined by the highest elevation of contours rising at the top of hill ranges. It becomes higher than the surrounding plane. The catchment area of the basins are demarcated by water divides. Figs.1.6 and 1.7 showing drainage basins and drainage orientation in the region clearly indicate the complex nature of water divides in the region.

(c) *Residual Isolated Hills, Erosional Remnants (Monadnocks), Rock Exposures, Domes etc.*

Very high dissection appeared in Pachmarhi plateau. Residual hills over the plateau rim due to lithological changes are most common throughout the region. Lava exposures of Pandav cave which is a isolated hill over the plateau, dissected isolated Ganesh hill, biologically weathered and consumed Ganesh hills by denudational processes, Pandav cave, hills in the northeast of Pandav cave and hills remnants near Nauka Vihar (Vanshri Vihar) clearly illustrate the examples of residual isolated hills and erosional remnants over the plateau.

Rocky knob is noted along the Nagdwari valley (55 J/6), near Merki (55 J/6), near Rorighot Barghat Pahar (55 J/7) in the north of Dhupgarh Pahar (55 J/7), Mahadev Pahar Bariam Pahar, Marodev Pahar, near Bharna village (55 J/7) and so many places of the region (Fig.4.1). Lava dome can be seen on Pandav cave hill while domal shape of hills structured by sandstone rocks are most common in some parts of Jatashanker hills.

(d) *Knickpoints*

Knickpoints are the breaks of slope in the long profile of a valley where a new curve of erosion, graded to a new sea level intersects an earlier. Knickpoints terminate the erosion surfaces and define the stages of cycle of erosion. The longitudinal profile of river Denwa and Sonbhadra denotes several breaks in slope. These breaks make easy the recognition of former erosion surfaces. It is remarkable that the knickpoints only can be seen where two or more cycle of erosion have been undergone. Thus, these knickpoints determine the formation of multicyclic landforms in the area.

(e) *Rills and Gullies*

Rills and gullies are the most common features of the area. These are found along the lower section of the scarps, valley sides and along the morrum accumulation of isolated hills. Most of the rectilinear slope of the hills which are sedimented by debris are the birth place of hills and gullies. Gullies and rills are parallel in nature from top to lower section of the slope profile.

Features of Valley Development

River valley incorporates assemblage of landforms. The nature and characteristics of valleys depend upon the dominance of fluvial processes. The recognition of landforms of valley development can be recognised through valley thalwegs.

Most of the river of the area form deep gorges and thus V shaped valleys are most common in the region. V shaped valley formation can be marked near Handikhoh Pahari where steep valley incorporating steep gorges in both sides are well developed. Meandering valley is recognised near Chaura Darshan. A view of V shaped valley formation is mirrored in the west of Dhupgarh Pahar. V shaped symmetrical deep valley along with developed gorges must be seen near Denwa Darshan along Pachmarhi-Pipariya road. Valley along with structural benches in both the river sides is the best example marked near Jatashanker hill and in Jambudweep valley. Steep valley gorges due to solution of spring river water can be seen Jatashanker Mandir. Near Handikhoh well developed V shaped valley is marked. U shaped valley is marked in the opening valley path of Jambudweep river near Ganesh hills. Asymmetrical valley is also marked near Chaura Darshan. Thus vertical erosion and deepening of river valleys are the most common features of the area.

Development of Waterfalls, Rapids, Springs, Seepages, Kunds and Khads

The development of waterfalls are possible where the river leaves the plateau rim while the rapids are found due to faulting or the exhumed rock exposures appeared due to fluvial processes. Springs and seepages belong to scarpline generally where the different type of rocks are interlocked each other. The cave position denotes the formation of springs and seepages.

(a) Waterfalls and Rapids

The existence of waterfalls and rapids make the riddle of the geomorphic history of the region more complex. The location of waterfalls and rapids is given in Fig.4.1. Pachmarhi plateau is well fascinated by waterfalls and rapids. A small rapids is noted as Nimboo Bhoj between Lashkariya cave and Jatashanker. A rapid is also seen near Jatashanker along the Jambudweep valley. Near Atake village located on Jatashanker hills, and along the Pachmarhi-Pipariya road near Bari Amlake rapids are also visible. In the way of 'Nagankhand' near Khamkheri village a 50 feet high waterfalls is located. Bee fall (Jamura fall) is about 300 feet high and marked in Madhuvan at the height of 1124 metres on the topographical sheets.

There is a chain of waterfalls near the township itself. In the east just by the side of the family quarters in the military area known, as Sanitarium walk just one km and reach Pavas Prapat. This waterfalls becomes a great fall in rainy season and water falling down from a height of 75 feet has no equivalent on beauty. One km ahead of this fall is another fall known as little fall. The height of the fall is nearly 7 feet. On the other hand other falls viz. Bee fall, Duchess must have aroused a feeling of uneasiness because of heavy downfall. Nearly 100 yards to the west of the little fall there is another fall known as Down fall. This is also a beautiful one and becomes more beautiful in rainy season. If you have courage, physical fitness, time and most importantly fondness for nature then walk downwards and you will come across more waterfalls one by one. Though it is impossible due to typical terrain.

Rapids and waterfalls can be observed near Apsara Vihar, Rajat Prapat is a big fall located downwards the Apsara Vihar. The waterfall near Apsara Vihar is 30 feet high and in the upper parts it shows step falls. The Rajat Prapat looks as if a strip of silver is gliding through. It is rather difficult to descend and tough to reach near the Rajat Prapat as it is nearly 3 km and is very strenuous. Only adventurous and trekkers can negotiate the path which goes round the hill on the left of the view point. A small waterfall is also noted near Chilakdhar at 5 km distance from Pachmarhi to Pipariya road. Kelakhad (Sundar Ban), Khad Jhiriya and Phishala Khad, Jaishri Prapat (Shishu valley 8 feet in height), Balshri waterfall near water meets (50 feet height), waterfall near Barkachhar village (15m) and along Silpali Nala (55 J/6) are the best examples of waterfalls. The surrounding areas of Pachmarhi plateau are thickly dotted with Rapids and waterfalls.

(b) Springs and Seepages

Natural outflow of groundwater on the surface is known as spring. There are numerous spring conditions and seepage sites (Fig.4.1). Springs and seepages are very common features throughout the region. These are easily seen near Pavas Prapat, Kanjighat, Jatashanker hill, Mahadev Pahar, Patarkota Pahar, Lathikhap Pahar, Kiwarkhap Pahar, in the southwest of Chauragarh Pahar, Ghughudev Pahar, near Silshill village, near Bhagoli village, Mahuldeo Pahar, Mankidev Pahar (55 J/17), near Barachhar village and near Singanama village (55 J/6). The springs provide very cold water. Water appears on the surface where the country rock is weathered. These springs are the major source of water supply in the plateau region because besides these water bodies there is no other sources of water for residing population. Seepages sites are most common where spring conditions are noticed.

(c) 'Kunds; 'Khads' and 'Khoh'

Kunds (pools) and khads (deep valleys or ravines) are most common features principally formed by fluvial erosion. Khoh is the features most similar to caves. Bari Am Jhil (3 km away from Pachmarhi) near Bari Am village, Longi

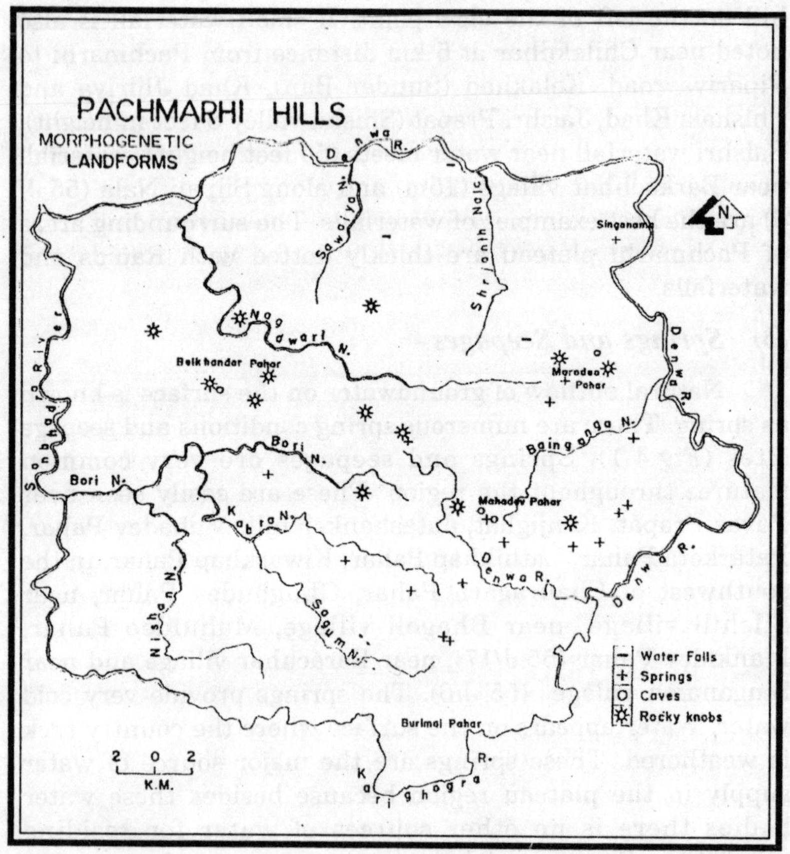
PACHMARHI HILLS
MORPHOGENETIC LANDFORMS
Denwa R.
Sowaiya N.
Shrijnath Nala
Singanama
N
Sonbhadra River
Nag dwari N.
Belkhandar Pahar
Moradeo Pahar
Denwa R.
Bori N.
Bori N.
Bainganga N.
Kabra N.
Mahadev Pahar
Denwa R.
Denwa R.
Nishan N.
Sagum N.
Water Falls
Springs
Caves
Rocky knobs
Burimai Pahar
Kalighagra R.
2 0 2
K.M.

Fig. 4.1

(Jhhela) Kund (located near Jhhela village above 3 km east of Pachmarhi Town), Ghoghara kund (Ghoda Awsani), Khad Jhhiriya kund (in the south of Gandhi Chawk along Patel road), Twaynam Pool (2 km away from Pachmarhi town), Sakharam Kund, Jamna Kund, Wellview Kund, Iron Kund (near Reechgarh), Sunders pool, Prabhu pool (over Dhupgarh valley near Shishu valley), Rajendra Kund, Kunds between Vanshri Vihar and Twidhara, Pansy Pool (Vanshri Vihar), Fairly Pool etc. are the beautiful kunds of Pachmari plateau.

Darmarra Khad, Sandawala Dhonga Khad, Nagan Khad, Denwa Khad, Kela Khad, Sakharam Khad, Fanshawe Khad, Daisy Khad, the Grevasse Khad, Phullar Khad, Panic pass Khad etc. are the Khads (deepest point) of the area.

Lashkariya Khoh, Hanuman Khoh (in the north of Gandhi Chawk), Jatashanker Khoh, Gangari Phool Khoh, Panarpani Khoh, Madadev Khoh (50 yards in length) Baniya Beri Khoh (50 yards in length) Khatkhata Khoh, Handi Khoh, Reechgarh Khoh (50 feet long). Astachal Khoh (200 yards in length containing rock painting), Dorothy Deep (Bhrantneer) Khoh, Nagdwari Khoh including Chintamani Gufa (100 feet in length), Swargdwar Khoh, Paschimdwar Khoh, Agamdwar Khoh and Chitrashala etc. are major khohs of the region. These are also known as Gupha or Caves. Nimboobhoj, Mahadeo, Baniyaberi, Ishanshring, Pandav cave, Mahadeo, Jambudweep, Mount Rosa, Dorothy Deep Fairly Pool, Adamgarh, Jhalai, Sonbhadra, Kajari and Bori cave or khoh represent rock painting in different colours from ancient period.

Depositional Landforms

Depositional landforms are most common in the valley region. Alluvial plain, fans and cones are notable features. Boulders controlled valley of Jambudweep near Jatashanker hills is the best examples of depositional landforms. Denwa Valley indicates different size of boulders. Alluvial cones are seen along the hill side slopes due to deposition of transported materials of different sized and shaped. Alluvial cones exist by erosion and weathering processes. Weathering processes form Talus cones due to mass translocation and deposition of rock wastes in the lower section of the hills. The system of sedimented materials differ in each condition. Erosional cones exhibit big boulders while weathering cones (Talus cones) show finer particles in the upper parts.

CONCLUSION

On the basis of the above discussion, it may be said that the region is well ornamented by the assemblage of landforms either produced by weathering processes or by fluvial processes. Denudational processes have caused a series of landforms. The landforms are changing their form through time and more or less controlled by existing lithology and processes. Waterfalls and breaks in slope illustrate that the region has experienced several tectonic jerks. The parallel retreat and downward movement of the slope and existing terrain clearly indicate the late mature stage of the region. A to Z beauty, performance and landforms assemblage recite the fine environmental circumstances in the region.

REFERENCES

1. Davis, W.M. 1930: Origin of Limestone Caverns, *Bulletin of Geological Society of America,* 41, pp.475-628.
2. Faniran, A. 1971a: Implication of Deep Weathering on the Location of Natural Resources, *Nig. Geol. J.* 14(1), pp.59-69.
3. Ollier, C.D. 1969: *Weathering,* Longman Group Limited, London.
4. Prasad, G. 1986: Morphogenesis of Guptagodavari Caves in Vindhyan Dolomite Rocks Near Chitrakut, M.P., India, *Modelling, Simulation and Control,* AMSE Press France, C, Vol.5, No.1, pp.31-42.
5. Prasad, G. 1986: Slope Evolution and Profile Characteristics of Chitrakut Upland, Proceeding of International AMSE Conference held in New Delhi.
6. Prasad, G. 1987: Morphogenesis and Description of Major Landforms of Chitrakut Upland, India, *AMSE Transactions,* AMSE Press, France, Vol.1, No.2, pp.23-44.

7. Prasad, G. 1991: Morphological Processes, Landforms and their Significance in Environmental Management of Chitrakut and its Adjoining Areas, Ph.D. Thesis (unpublished) Submitted to Kanpur University, Kanpur.

8. Singh, S. 1977: Evolution of Granito-gneissic Tors and Cutt-off Spur Mounds of Ranchi Plateau, *National Geographer*, Vol.12, No.1, pp.93-97.

Tourism Development

HISTORY OF TOURISM DEVELOPMENT

Pachmarhi is Madhya Pradesh's most verdant jewel. It is a beautiful shelter where nature has secured exquisite expression in myriad enchanting ways. If on earth be an eden of bliss, it is this and none but this. This saying is the most befitting for Pachmarhi—a hill resort and rightly called the queen of Satpura. Showing glamour, beauty and pinching scenic sensation, this hill station is a religious place for the Hindus as lakhs of people visit this place to pay their homage to lord Shiva on 'Shiva Ratri' and 'Nag Panchmi' and on other auspicious days. As a trekker's place, Pachmarhi is the Trekkers paradise and has no equivalent for it has the great vistas of hilly tracks to trek on. This hill station is the shelter of peace, affection, ebullition, elation, emancipation, emotion, engagement enjoyment, entertainment, excursion and excitement. It is also the most ideal place for health and happiness and thought to be as the happy home of excitement for the honey-mooning couples where they can find lovely places and can be all alone except the silent hills, pleasant winds and the melodious singing birds.

The most attraction of deep valleys, ravines, narrow cut deep gorges sculpted in red sandstone by the running water, wind and weather, multi-coloured forests, long winding roads and streams full of limpid cool water, the little charming rapids and great cascading waterfalls, hanging valleys and hills showing mesas and buttes, smiling gardens and silent scene of rich vegetal covers surrounding the Pachmarhi hills attract the peoples come to meet here.

Pachmarhi presents a fine gift of topo-culture where the setting sun colours the green leaves by its rays, scattered hills standing in silent prayers to the power unseen, waterfalls giving an allusion of bees and buzzing, deep ravines and valleys to give us how deep depth could be, dark silent caves where we could feel the coolness of silence and rediscover ourself.

The original name of Pachmarhi was 'Panchmarhi' based on Panch Pandavas which can be traced back to the period of Mahabharat. Historical evidences advocate that Panch Pandavas have passed their exile period in the Pandav caves of Pachmarhi. It is believed that Panch Pandavas have spent thirteen years of the exile periods in Sohagpur tahsil of Hoshangabad district. The twelfth year passed in Sandia village of Sohagpur tahsil while the next thirteenth year of exile was lived by Pandavas in Pandav caves. Thus it is clear that the Pachmarhi is the oldest shelter for the tourist as it is named and famed by Pandavas since Mahabharat times.

The modern history of tourism development in Pachmarhi does not go far behind. In 1862, Captain Forsyth J. reported about the sylban-charms of the place to the chief commissioner (Sir Richard Temple) of Hoshangabad. He was so much impressed that the Captain was asked to go Pachmarhi again to examine the possibility of using the plateau for establishing a sanitarium. On receiving the positive response a sanitarium was built which is now turned into family quarters for military personnel. However, the first building in view of tourist encroachment in Pachmarhi that was constructed was named as Bison lodge which is now a centre of attraction for tourists as it is museum called 'VANIKI SANGRAHALAYA'.

The Pachmarhi as a whole was previously under the rule of 'KORKU JAGIRDAR'. It is reported that this pleasant area was much attracted to military men and British rulers. A military centre was established here in 1870 and after seven years a local body government, the Contonment Board existed. Pachmarhi was too precious to the British. They did not permit general public to see it. Even the mates and coolies who were employed for construction work at Pachmarhi were instructed to keep their colony on the foothills presently called Matkuli. The English did not want the town to grow into a city of affuence destroying the calm and repose so the rules were framed in such a way that the construction of big houses or two storey buildings higher than the prescribed limit was not possible. Even the Nawab of Bhopal who used to come here frequently to play POLO was not permitted to construct a Palace in the area of the contonment and he could get the Palace built on the outskirts of the township. The Palace was later purchased by the rulers of Nasirgarh. The ruins can still be seen at the doorship of Pachmarhi (Bansal, 1991). Thus it is clear that Pachmarhi was not used as a public place or as a tourist centre during the British period.

As regards the area development, the Pachmarhi contonment area was inhabited by the Indians who were serving as the British officers of the army or other Gora Sahibs. In course of the time the cantt. developed into a town in 1931. The other parts of Pachmarhi is now under the administration of the Special Area Development Authority (SADA) which was established in 1976.

The hill station of Pachmarhi was not marked on the map of All India tourism till recently but many great Indians had stayed here. The first and foremost comes the name of Dr. Rajendra Prasad, the first President of free India. He was so fascinated by the beauty of the place that he visited Pachmarhi twice and stayed here for several days. Dr. Lohia, Sri Jai Prakash Narayan, J.B. Kripalani, Ashok Mehta are the few names who had taken abode here for recouping their lost energy. The first commander-in-chief General Cariappa had also worked here as an officer in the Army. Dhyanchand had served as an N.C.O. here.

People from film industry, though are very much fascinated by the untouched beautiful scenery and would like to shoot as many films as possible but the time involved in the journey to and fro makes it rather difficult for them to come here with all the paraphernalia essential for shooting a film besides bearing the expenses that would occur on the transport. Nonetheless, a few films viz. Quismat, Pal Do Pal ka Sath, Dil ne Ekrar Kiya, Massey Sahib, Thoda sa Rumani ho Jayen and the Electric Moon (A film for channel 4 of the B.B.C.) have been filmed here.

The cave paintings in Pachmarhi also illustrate the oldest story of the area. The Mahadeo cave is nearly 300 ft. long. Most of the paintings have been placed in the period 500-800 A.D. but the earliest paintings are estimated 10,000 years old. The Dorothy rock shelters are believed to be of 10 to 15 century B.C. They depict the scenes of hunting, dancing, and war and some of them are fully coloured. The rock painting near Rajat Prapat is believed to be of the 4th century B.C. These historical evidences advocate the tribal encroachment in the early period of the history.

It is clear by previous records that the road facility and nominal transportation was started over the Pachmarhi hill in 1870 by the establishment of military centre and by expending Rs. 50,000 at that time. The tourist encroachment has been frequently noticed after independence while before the independence, it was totally governed by British rulers and the tribal population existed mainly in three Jagirs i.e. Chhater, Bariam Pagara and Pachmarhi. Now the area is opened for all peoples coming from different parts of India and abroad but the permanent inhabitation is restricted beyond the local rules.

SITES OF TOURIST'S INTEREST

The most attractive places for the tourists are marked in FIg.1.10. After travelling 54 km road distance between Pipariya and Pachmarhi, a beautiful scene of Denwa Darshan make a strange for tourists due to deepest valley of Denwa draining in southeast direction located about 8 km upward from Matkuli. The valley is about 500 feet deep from the running road surface and gives silver shining of moving water during sun and moon light. There are more than hundred sites of tourists interest which may be settled in 10 groups (Jayaswal, 1952).

The Interesting Sites of Jatashanker Group

(a) *Hanuman Cave:* Hanuman cave is located along Pipariya road in the north of Gandhi chawk. Hanuman cave is just like a shelter.

(b) *Lashkariya Khoh:* It is two furlongs away from Hanuman cave. A cave is also located here. D.H. Garden has stated this cave as Bazar cave. The hills near the Bazar cave is pointed by tribal population of First century B.C.

(c) *Nimboo Bhoj:* Moving forward the Lashkariya khoh in the way of Jatashanker, there is a spring draining over the calender year is known as 'Nimboo Khad Ki Jhiriya'. Nimboo Bhoj is a beautiful place covered with Nimboo trees and surrounded by red coloured painted hills.

(d) *Jatashanker:* Jatashanker is a temple fashioned by the creator's eye. It is situated only 1.5 km from bus stand of Pachmarhi.

Jatashanker is a place of abiding sanctity and is the best gift of nature presented to man in the most undiluted form. The samadhistha SHIVA under a cool dark cave is one of the myriad manifestations of nature inviting man to come closer to her. You can take a dip in the cool water of Jambu river which takes its source from here. The overcasting little rocks, many of them hanging precariously between the huge walls of rock cutting are a thing of wonder. The calm and repose here would assuage a person who is mentally and physically tired and would certainly draw new inspiration from its sublime natural beauty. The mass on the rocks around the little respository is of a special value. It changes colour from green to golden yellow due to light reflection. If you have seen the film 'PAL DO PAL KA SAATH' or 'Massey Sahib' you might be familiar with the scenery. On the way there is a temple of Lord Hanuman. This idol has been carved out by a local craftman. The temple has been constructed by public donations. Even the ex-Governor of M.P. state Hon'ble Sri S.N. Sinha also contributed money for the beautification of the temple.

(e) *Atakey:* Atakey is situated 1 km away from Ghoranagar village in the way of Jatashanker. The place is famed by magnificent spring between mango trees.

The Interesting Sites of Rock Doh Group

The interesting places of Rock Doh group can be seen by using motor cycle. It includes the following sites of tourists interest.

(a) *Siriyan Valley:* Siriyan valley is located after 31 km passing on the Pipariya road from Pachmarhi. One can see the Dhupgarh and Chauragarh hills from this place.

(b) *Bari Aam Lake:* Bari Aam Lake is located 3 km away from Pachmarhi. 'The water stored in the lake is bluesh in colour. The surrounded area is covered with the huts of Gond tribes. Bari Aam village is also famous by the oldest Jagir of tribal population namely as 'Bars Aam Pagara'.

(c) *Darmarra Nala:* This site is 30 km away from Pachmarhi. It is a picnic spot surrounded by hills dense forest, moving water and almost plain and rolling surfaces.

(d) *Chilakdhar:* Chilakdhar is a drain about 29 km away on Pipariya road from Pachmarhi. It is also beautiful place for picnic.

(e) *Sandawala Dhonga:* It is a silent hill centre about 29 km away from Pachmarhi on Pipariya road. The panoramic view of Pachmarhi can be observed from this site.

(f) *Khirkhiri Pahar:* It is just like a mesa hill located 28 km away from Pachmarhi. The upper part of the hill is like a table land.

(g) *Nagan Khad:* Nagan Khad is located about 29 km away from Pachmarhi near Khhamkheri village. A beautiful waterfall is visible here where the existing river moves making meandering path and may be treated as 'Nagan'. The area is well dissected by river actions.

(h) *Panar Pani:* It is a picnic spot 28 km away from Pachmarhi covered with mango and jamun trees. The draining nala makes the situation beautiful.

(i) *Kuppi Lane:* It is also known as 'Gangari Phool'. It is religious place. Swami Biththal Deo ji Brahmachari, and Swami Sri Prakashji have passed their life at Kuppi Lane. The spring condition at Kuppi Lane provides picnic facility.

(j) *Dana Nad:* It is located near Pagara village about 8 km away from Pachmarhi. Spring condition, flowing nala, wide plain and charming forest with hilly tract attract the site of Dananad.

(k) *Harra Kot:* It is the place historically known as the capital place of Korku jagirdar's king Vibhuti Singh situated in the north-west of Dananad. 'Phasi Khad' of Vibhuti Singh is also located here.

(l) *Denwa Khad:* One can observed the Denwa Khad marching 10 km away from Pachmarhi along Pachmarhi-Pipariya road. You can see here the zig-zag path of Pachmarhi-Pipariya road, the beautiful scene of Narmada valley in a long range, the greenish belt of lower Denwa valley and the thick blanketing of Satpura forest. This site is also termed as 'Washan valley'.

(m) *Rockdoh:* It is located near Singanama village. A Rest House is constructed near Rockdoh for the purpose of tourists. It is a beautiful place for catching fish in silent Denwa river and suitable for picnic festival. Denwa river forms various blue water pools near Rockdoh which is famed as 'Tiran Dahar' and 'Rani Dahar'.

(n) *Langi (Jhela):* It is located near Jhela village. The river Denwa flows here very slowly. The river follows several kunds in their valley surrounded by high hills.

(o) *Ghora Ausari:* It is one km away from Ghoghara kund and located in front of Jhela. It is a mesa like flat top hill where the upper part is dotted with horse feet signs.

The Interesting Sites of the Catacambs Group

(a) *Marhadeo:* Marhadeo hill is located on Pachmarhi Pipariya road. The hill is famous for 20 m long cave.

(b) *Baniya Beri:* Baniya Beri is famous for its 50 mts long cave and oldest rock paintings.

(c) *Sambourne Cave or Marhadeo Cave and Rock Painting:* This shelter is located on the old road to Pipariya. There are a group of shelters and the nearest to the place, one can leave their vehicle is Chieftain's cave. The Marhadeo cave is nearly 300 ft. long and you will certainly like the mysterious silence pervading there. The rock paintings of Marhadeo are believed to be of very ancient period.

(d) *Mayhew Peep:* A beautiful hills located in the eastward of the city. The hills cover some rock painting in its south-eastern flank.

(e) *Pavas Prapat:* It is about 3 km away from Gandhi Chawk. There is a chain of waterfalls near the township itself. In the east just by the side the family quarters in the Military area known as sanatarium walk just one km and reach Pavas Prapat. This waterfall becomes a great fall in rainy season and water falling down from a height of 75 ft. has no equivalent on beauty. It is about 300 ft. long and 90 to 120 ft. in width in their lower section. There is a pool also where you can go for swimming. In summer days this fall returns into a remblance of a fall.

(f) *Down Fall and Little Fall:* One km ahead of Pavas Prapat is another fall known as Little Fall. The height of this fall is nearly 7 ft. and bathing under this fall has its own charm and pleasure because on both sides of the fall small rocks make a natural bathroom. The height is not much so while taking a bath you do not feet any sort of uneasiness irrespective of the duration under the fall. Nearly 100 yards to the west of the Little Fall, there is another waterfall known as Down Fall. This is also a beautiful one and becomes more beautiful in rainy season.

(g) *Sundar Van (Kela Khad):* It is located near Down Fall, It is famous for waterfalls, Kunds and for Kadali Van.

(h) *Kitty Caves:* Kitty caves is located in a hill found near May Hew Peep about 5 km away from Pachmarhi. The view of Pachmarhi can be seen from this hill.

(i) *Bawanganga Kajighat:* It is located about 5 km eastward of the Pachmarhi town. The draining small tributary near Kajighat village associates beautiful scene for tourists.

(j) *Khad Jhiriya:* The place is also known as 'Phisala Khad' where a rapid of 7 ft high attracts the people beside washing ghat and fish catching ghat along the drain.

The Interesting Sites of Chauragarh Group

(a) *Pandav Caves:* Pandav caves are located in the south of the city about 1.5 km away from Gandhi Chawk. Pachmarhi takes its name from these five caves carved out of a single hillock. It is believed that the Pandavas stayed here for quite a long period during their exile for thirteen years. The cleanest most airy of them is known as Draupadi Kuti (Hut) and the dark one, the Bhim Kothari (Small room). Archealogists claim that these caves must have been constructed by Buddhist monks in the 9th or 10th century A.D. but the popular belief that the Pandavas had lived here, still continues.

(b) *Handi Khoh:* Handi Khoh is a deep and large khud. It is nearly 300 ft deep. Try to throw a pebble to the bottom and surely you will fail in your attempt. People believe that the Lord Shiva killed a great snake which actually was a demon and burried it in this khud. The local natives used to call it Andhi Kho which with the time got deshaped and now is known as Handi Khoh. Handi Khoh is made by Handi Khoh Nala which takes their birth near the town.

(c) *Mount Morris:* It is a small hill near the Handi Khoh and locally termed as 'Akant Giri'. The panoramic view of Pachmarhi can be observed from this place.

(d) *Malcom Point:* It is located in the way of Handi Khoh. One can easily see the view of Chauragarh and the associating valleys from this point.

(e) *Mayne Rock:* Mayne Rock is also known as 'Azad Shringa'. From Handikhoh this hillock is nearly 2 km. From the top of the 'Azad Shringa' you can have a grand view of Mahadeo and Chauragarh hills in the south and of Dhupgarh in the west. The allround scenery is enhanting.

(f) *The Bishap Spueeze:* This point is located downward direction of Chaura Darshan. It is named as 'Giri Darar'.

(g) *Island Rock:* It is a small hill which has more than 100 acres beautiful plainland covered with forest over the top surface. It is located near Chaura Darshan.

(h) *Forsyth Point:* It is located on Mahadeo road and about 5 km away from Pachmarhi. It is the place from where you can easily see the beautiful scene of Chauragarh and 'Azad Shringa'. It is also known as 'Priyadarshini'.

(l) *Bara Mahadeo:* Regarded as holy for countless generations, Mahadeo hill has a shrine with an idol of Lord Shiva and an impressive Shivlinga. On the east side of the hill is an excellent cave shelter with paintings. The inside of the cave is so cool that the coolness inside can be experienced even from outside. As you enter the cave you enter into a world of tranquility and religiousness. You can have a dip in the holy depository. This depository is formed by drops of water dripping from the roof of the cave. The all prevading peace makes you feel closer to yourself and to the supreme. The idol of Shiva was installed by the Mahadeo Mela Samiti. The original Shivalingam is also beside this idol. It is believed that Lord Shiva did a 'Tap' in the cave after destruction of Bhasmasur. The demon was lured by Lord Vishnu who had taken form of a beautiful damsel to put his hand on his own head which resulted in his destruction. The pond outside of cave is known as Bhasmasur Kund where the demon was actually burnt to death. Mahadeo has been a centre of pilgrimage for the last many centuries, since time immemorial. Hindus has been coming to Mahadeo to pay their homage to Lord Shiva. In modern days a large fair is held on Shiva Ratri. Nearly two to three lakhs of people visit Mahadeo and Chauragarh on this occasion. People who have their wishes fulfilled come to pay their gratitude while others come to invoke the blessings of the Lord.

(j) *Gupta Mahadeo:* This place is a place of wonder of nature and revered as a sacred spot. This is a narrow point in the valley with rocks overhanging a stream and a small spring from which water cascades down. Enter a long, nearly 30 ft. narrow dark gully in which you can barely pass through, you can have darshan of a natural Shivalingam. An exhaust fan fixed by local authorities at the entrance enables you to breathe easily.

(k) *Chauragarh:* It is nearly 3 km from Mahadeo. It is one of the Satpura's prominent land marks, and the sacred summit is crowned with emblems of Mahadeo worship. The path of Chauragarh is strenuous and you have to climb nearly 1300 steps. Only a fascination for natural scenery intermingled with devoutness will enable you to reach the top and once when you reach the rectangular top, cool winds will freshen you and you will be charmed by all around you. The top is nearly 240 metres long and wide. An idol of Lord Shiva has been installed by the devotees. A temple is also constructed.

This hill top has been a centre of Pilgrimage for the Hindus who have been coming to offer their prayers to Lord Shiva for the last many centuries. A tradition that you may not find anywhere is the offering of Tiscends (Trishuls) of different weights as a mark of devotion. Some of the Triscends must be weighing three to four quintals. To carry such heavy weights on to such a height is possible only when devotion transcends strains of physical labour. In these days when religious beliefs are under questions you can still see several pilgrims at the time of Shivaratri fair carrying triscends on their shoulders and climbing the hill non-stop. There is also a Dharmshala where you can relax a while. The M.P. Government has got the place electrified. Drinking water is also available here which is lifted up from the tubewell at Mahadeo.

The Interesting Sites of Jambudweep Group

(a) *Twaynum Pool and Jambudweep:* A pool that provides facilities for swimming to all type of swimmers is Twaynum Pool (RAJYAPAL SUR). This pool was constructed at the behest of the last British Governor Sir J. Twynum. A natural stream supplies water to the pool. The pool is safe enough for children and non-swimmers too. One can enjoy a good bath and swimming here. The pool is maintained by the SADA. If you walk downwards another 2 km from this pool you will reach an open plain. This is Jambudweep. From Jambudweep you can proceed for Chanagarh where there is rock shelter consisting of rock paintings. The rock shelter of Jambudweep is about 100 feet in length.

(b) *Chhota Mahadeo:* It is situated in the west of 'Rajyapal Sur' about 3 km away from the town. A rock shelter may be traced here in a small hill. The rock shelter contains oldest rock painting in white colour. The spring condition is also noticed here.

The Interesting Sites of Reechhgarh Group

(a) *Raj Giri:* It is also known as 'Club Hill' and located in the backside of Pachmarhi Club eastward of Gandhi Chawk. It is 300 feet from local ground surface and about 3790 feet high from the sea level. It has been throwing a standing open invitation to persons who wish to beyond an allround bird view of Pachmarhi besides watching the sun setting behind Dhupgarh. The path to the foot-hill is near the Pachmarhi Club. This hill is not very difficult to climb. When you reach the top you can have a view of the township and landscapes allround. The silent tops of various hills appear to be in deep meditation. It will be interesting to locate your lodging place from that height. View of the sunset will also be yet another experience. After the sunset you come down the hill in twilight without any great difficulty.

(b) *Lamji Hill:* It is located near about Government House. It is about 3864 feet high from sea level. One can reach at the top of the hill by approaching the path in east and west directions. A beautiful scene of forest cover can be seen from the hills top.

(c) *Machhali Ka Mor Hill:* It is situated 2 miles away from the town and stands between Bee drain and Jambudeep river. Approaching the top from eastward direction one can traced the panoramic view of the forest.

(d) *Sakharam Khud:* It is a swimming pool located in Khud region westward of Governor's house.

(e) *Bee Dam:* Bee Dam incorporates a swimming pool known as 'Jamuna Kund' used for the swimming by the members of Pachmarhi Club. The swimming pool is fascinated by two dressing rooms. Due to accumulation of dirty water of the city it is now out of use. It is located in the upward direction of Bee fall.

(f) *Bee Fall:* This fall is a spectacular fall in the stream which provides drinking waters to Pachmarhi and only 3 km from the township. The road is nearly good and if you wish to walk on foot, it will prove to be a good idea. If you do not let your enthusiasm lessen, you should not miss a chance to bathe in the stream. Negotiate the downward path for half a km and reach the fall. Try togo under fall and enjoy the water beating on your body. It would not take more than five minutes to freshen you and very possibly you would not be able to stand the cold even in the summer. This is a place where a day can be spent gleefully with family and friends. Truly speaking one must bathe under the fall to have the real feel of it. The fall exists from 150 feet high scarp.

(g) *Fanshawe Khud:* It is above 3 km from the town and located in short circle in the way of Dhupgarh.

(h) *Bee View:* It is the point showing best scene of topo-culture.

(i) *Blundell's Bluff:* It is a hill about 2.5 km from the town. It is 50 yards away from Reechhgarh. One can see the Dhupgarh from the top of the hill. This memorial was raised in memory of the son and son-in-law of the then British Governor. Opposite to this grand memorial, stands Dhupgarh in a challenging posture but actually is easily vulnerable. This peak is the highest point in Satpura ranges and the highest point look like an elephant and is called so. The smaller one is called the tiger and the smallest one in between is the goat. It is also known as VATSALYA.

(j) *Reechhgarh:* It is a hill where a wonderful natural amphitheatre in the rock approached through a cave like entrance on the south side. It contains the cave of 50m in length. The distance from the town is 3 km.

(k) *Irene Pool:* It is locally known as 'RAMYAKUND'. If you have stayed for night at the Rest House on Dhupgarh, then after having watched a beautiful and enchanting view of the sunrise you would like to go to a place where you can enjoy bathing. This ideal place for the purpose is Ramyakund. On the journey back from Dhupgarh, before Reechhgarh, you see the plaque showing the path to the pool. Walk nearly two km and you reach Ramyakund. This pool formerly known as Irene Pool is a discovery of a lady Mrs. Irene and was named after her. This little pool is not deep but ripping blue and green. Clean water of the pool would surely entice you to have a dip. You will find three deep pools full of cool water. It is a picnic spot.

(l) *Monte Rosa:* This hill previously called Monte (Mount) Rosa is a lost spot for the modern tourists though it is very beautiful. The British lovers of nature had railing fixed for the help of the climbers. A view of the sun setting behind the Satpura ranges is one of the most immemorable views. There are four shelters with paintings, comparatively early linear drawings. Along the northern side of Jambudeep valley are some six shelters with many paintings of animals and human figures, indicating a detailed battle scene. This point is also known as Astachal.

(m) *Woodburn Cliff:* It is also known as 'Ramyak' located beside 'Ramyakund'. The draining rivers, the hills top of Dhupgarh in the west and the Jambudeep mountains in the north can be seen easily from this point. The deep valley between Ramyakund and Dhupgarh is notable.

(n) *Daisy Khud:* It is located between 'Astachal' and 'Do crag'. One can see the natural beauty of the hill ranges sitting upon the bench settled here.

(o) *Docrag (Durgam Giri):* It is located in the west of Reechgarh and in the south of 'Astachal'. It is a typical hill but easy to reach at the top for observing the panoramic view of hills.

(p) *Dorothy Deep (Bhrantneer):* Dorothy deep is defined by rock shelters. These rock shelters are believed to be of 10 to 15 century B.C. They depict the scenes of hunting, dancing, and war and some of them are fully coloured. In an excavation carried out years back a skeleton of a human being measuring nearly seven feet was found. Besides these rock paintings place has a natural setting and a typical charm also. Come back to the main path and proceed further. Covering a distance of nearly 1 km we reach near a brook. Walk downwards along that symphony of nature and you are at a point where this brook changes into a fall. The darkness due to the two huge gorges blackens the water and the fall becomes a picture of beauty.

(q) *Duchess Falls:* It is also known as 'JALAVATARAN'. The descent is steep and the trek strenuous for almost all of the 4 km to the base of the fall's first cascade. Please descend very carefully and after a km you come to the fall. This fall appears to be beaming with life and when you bathe under the fall you shed off all tiredness and fatigue. It is rather difficult to stay under the fall for more than a few minutes as you would begin feeling cold. Climb unwards and there you can have a view of one other grand fall. It needs much more courage, physical strength and skill to reach near it and one should avoid this adventure.

(r) *Sunder Kund:* From Duchess Fall you can reach Sunder Kund after a 35 minutes walk covering a distance of 2 km of a strenuous path. It is a pool of pellucid water too deep to fathom by any ordinary man. You are advised not to dare unnecessarily unless you are a very good swimmer.

The Interesting Sites of Dhupgarh Group

(a) *Government Ornamental Garden:* Spend your time (morning and evening) pleasantly without doing hard walking by going to the Government Garden. The lush green lawns be a treat to your mind and body. It would be certainly an unforgettable evening as flowers emitting a swan of scent and a poignant sweet fragrance would match your mood. You would surely wish to spend your evening here for the remaining days you plan to stay at Pachmarhi.

(b) *Mrignayani:* A unique place that is meant for animals who get wounded in the forest and might succumb to their wounds. People from sanctuary bring such type of animals and keep them here. They are treated by the veterinary Doctor and when they get all right they are left in the forest to led healthy and free life. These animals are Deer, Cheetal, Sambhar, Neel Gai etc.

(c) *Van Vihar:* By the side of the Government Garden, children have plenty of playing facilities and amusement waiting for them. Let them go there and enjoy their sojourn. The SADA has set a children's Train in 1992 in this Van Vihar.

(d) *Sidoli:* By the side of Amaltash in Pachmarhi urban area, you find a collection of some stone and wooden sculptures heaped under a big mango tree. This is SIDOLI, a cultural heritage of the Korkus who are the aboriginals of this area. For every dead a stone or wooden inscription is placed to satisfy the departed soul. The script is illegible and is beyond comprehension.

(e) *Polo Ground Garden:* Going towards Dhupgarh from Champak, you will find the Polo Ground Garden, Horticultural station of M.P. Government is also established here. It is now used for the plantation of fruit trees.

(f) *Keating Point:* It is also known as 'Sangam Dur'. It is an attractive place that can be reached by descending a meandering path nearly two km long from a point at the long Chukkar Road. Two streams of water meet the slim Denwa river which has its origin near Dhupgarh. Surprisingly water here is not so cold in winter as one might expect while in summer, it is real cool. Here you can bathe and swim. From here you can swim in the watery gully for a long distance and can reach Agam Triveni. You definitely require a life jacket and company of good swimmers. Though now usually nobody takes the risk and so you should also not indulge in this type of adventure.

(g) *Grump Crag:* It is locally called as 'Sushmashar'. It is situated between 'Sangam Dur' and 'Van Vihar'. It is a beautiful place of attraction on the long 'Chukkar Road'.

(h) *Giri Darshan (Lady Robertson's View):* It is situated near Vanshree Vihar. You can see the highest peak of Satpura from this point.

(i) *Dhupgarh:* Dhupgarh is the highest point in the Satpura range with a magnificent view of the surrounding ranges. It is a very popular spot for viewing sunset. The peak of Dhupgarh, the so-called Elephant is 4429 ft. high from the sea level. It was known as Shri Harvatsakot in the olden days. The central railway has established a micro-wave station here. The allround beautiful landscapes, vehicles looking like toys below on the road and different peaks of the range would certainly enchant you. The moments will leave a deep imprints on your memory. The sun, sliding slowly on the horizon for its nest, colours clouds with different shades with a golden bush and presents to the world a kaleidoscope of colours. While the whole forest with all its inhabitants bids a silent farewell to the life giving sun; evening begins stretching its shadowed arms blanketing the earth darkness. There then we come out of the divinely experience not experienced before and enter this earthly world. To remain consciously for a few moments in a place like this will be an experience that transcends a human mind can think of what one seeks and finds here is not only pleasure but an inner harmony the perfect tuning with the infinity. A silent you might enter the world of tranquility where the beating of the heart can be heard. Here you would feel that silence has also a sound no ear can hear but feel. You will also appreciate adventurist nature of the British who built a Rest House on the hill where one can stay for night and if luck favours can listen to the roaring of a tiger or leopard. Booking for the Rest House is done at the local P.W.D. office.

(j) *Hog's Back:* It is also known as 'Ballabh Prasth' located in the way of Dhupgarh. It is 3678 feet in height.

(k) *Fraser Gully:* If you walk along the oldest route of Dhupgarh you will find a nala draining all over the time where freezing water quenches the thrust of the tourists. It is a beautiful picnic spot.

(1) *Childrens Valley:* It is 1 km away from 'Jalgali' (Fraser Gully) in Dhupgarh valley. It is a picnic spot.

(m) *Prabhu Pool:* It is located near childrens valley existed in Dhupgarh valley region. It contains a swimming pool.

(n) *Rajendra Kund:* It is also a swimming pool located near Prabhu Jal Kund.

(o) *Jai Shri Prapat:* The nala located in childrens valley forms a 12 m high waterfall which is locally termed as Jai Shri Prapat. You can take bathe here easily.

The Interesting Sites of Rajendra Giri Group

(a) *Rajendra Giri:* This hill is fascinated by late Dr. Rajendra Prasad, the first President of India and is named after him. The Banyan that he planted has now grown into a tree. He visited this place a couple of times. A very important person's beautiful house known as Ravi Shankar Bhawan was constructed for his comfortable stay. A glimpse of the building can be had from this hill. The beautiful scenery allround will enchant you, particularly the three dominating peaks on the southwestern crest of the valley.

(b) *Fleet Wood Junction (Agam Triveni):* You can reach 'Van Shri Vihar' by another route also and that route is from Tridhara but only daring swimmers can do this feat. Descend down the road, the road just from the foothill of Dhupgarh and you reach Denwa Darshan. From there you can swim and wade through rivulet and river 'Van Shri Vihar'. On the route you will find many varied shaped pools and the total distance is 15 km. The rivulet flows down between two huge rock and is simply magnificent. Now something about difficult but one of the most charming places. It is of course, Nagdwar and other related dwars.

(c) *Bal Shri Jal Prapat:* Approaching the Agam Triveni you will find a waterfall of 50 feet high which is known as 'Bal Shri Jal Prapat'.

(d) *White Plash Gorge:* In the way of 'Tridhara' you will find the confluence of two nalas which is known as 'Ferm Bell'. Moving forward you will see the narrow gorge position known as Pot Hote. White fishes are very common there. This narrow deep gorge is named as White Plash Gorge.

(e) *Pansy Pool (Van Shri Vihar):* It is about 5 km from the town. One can reach here by water ways after 1 km swimming in narrow channel from 'Tridhara'. The place contains a big and beautiful 'Jal Kund'. It is a beautiful picnic spot but typical to reach here. For Botanical purpose, it is important place because the deep tool is surrounded by rare herbs.

(f) *Patharchata:* From Van Shri Vihar one can proceed for Patharchata trekking, wading and swimming in the stream. The name Patharchata means licking stones. The name seems funny but when you see the gorge flanked by unsurmountable huge rocky walls rising straight over your head and forget the slippery stone bed beneath the stream you are sure to fall in the shallow waters but very possibly get physically hurt. These huge walls proclaim unique rock formation of Pachmarhi plateau. This grand view is reserved only for those who have not only unflinching love for nature but physical strength as well (Bansal, 1991).

(g) *Waters Meet:* It is also known as 'Phullar Khud'. Two small tributaries meet here with Denwa river and the confluence point is termed as 'Sangam'. It is a typical point.

(h) *Titanga Pahar:* One can reach at this point from Van Shri Vihar. It contains high peaks.

(i) *The Chimney:* To reach at waters meet, it is a approaching route in the western valley side of Satpura.

(j) *Marten's Leap:* It is also a approaching route for waters meet full of lubricating rocks and typical in nature.

The Interesting Sites of Apsara Vihar Group

(a) *Clematis Point:* It is a hilly point back side of Military Hospital and covered with 'Khinni trees about 5 km from the town.

(b) *The Green Patch:* It is a beautiful natural ground covered with 'Durva' grass. One can reach here by approaching the route beside officers mess of Military campus.

(c) *Panic Pass:* It is a high hill in the way of Apsara Vihar.

(d) *Collection Crag:* It is about 2.5 km away from the Pachmarhi town in the way of Apsara Vihar. One can see from here the Bainganga valley in the east.

(e) *Apsara Vihar with Rock Paintings:* Apsara Vihar is a fairy pool. By the side of' the Pandava caves a road leads us to this spot. The pool is formed by a little fall which is nearly 30 ft. high. The pool is an ideal place for swimming and diving. A swimmer can reach under the fall after swimming the short distance of the pool and can have a good bath under the fall. But little ones and non-swimmers need not be disheartened. They can enjoy their dips in the shallow water which gradually gets deeper near the fall. Beside the Apsara Vihar, if you have a guide you can walk to see the rock paintings. This rock shelter is nearly 150 ft. long and has white and red coloured paintings believed to be of the 9th century B.C. A friendly advice—please do not try to find this rock shelter without the help of a guide lest you should miss the path.

(f) *Rajat Prapat (Big Fall):* About half a km towards the east of Apsara Vihar, you come a place from where you have a glimpse of the grand fall known as Rajat Prapat or Big Fall. The height of the fall is nearly 350 ft. and it looks as a strip of silver is gliding through. In the vast vista of solitude a flying birds present loveliness personified. It is rather difficult to descend and tough to reach near the fall as it is nearly 3 km and is very strenuous. Only adventurous and trekkers can negotiate the path which goes round the hill on the left of the viewpoint.

The Other Interesting Sites of Satpura Hills

(a) *Black Stone Mangara:* About 6 km west of Matkuli village which is now a urban centre, village Mangara is situated. River Denwa drains along with black stones 1 km away from this village. The two water pools between the black stones fascinate the view of river Denwa and make the site as a picnic spot covered with dense forest patching.

(b) *Paras Pani:* About 9 km going on Pipariya-Pachmarhi road one can see the board of Paras Pani along the Jhiriya Naka. Paras Pani presents the beautiful scene of Denwa river.

(c) *Nagdwari:* Nagdwari, Chintamani and Chitrashala group of rock shelters are not only the most magnificent rock shelters of Pachmarhi but most religious places as well. Thousands of people visit these places during 'Nagpanchami' every year. Just as the ascent to Dhupgarh begins a footpath bifurcates for Kajri. This village is nearly

12 km from this bifurcation. This village is a halting place for the pilgrims. From Kajri people go to 'Nagdwar'. This place is one of the most sacred places for the Hindus specially of Vidharbha region of Maharashtra.

Nagdwar is a gully narrowed by two big rocks. The gully is 100 yards long and at its end there is an idol of NAGDEV natural of course. After paying homage to the Lord pilgrims go to CHINTAMANI and on the way visit AGAM DWAR, PASCHIM DWAR and other dwars like SWARGDWARS. These so-called dwars are huge formations of rocks which are umbrella like in shape and can shelter thousands of people. Beautiful ancient coloured rock paintings are clear and are most precious for Archaeologists. Then there is again a halting place CHINTAMANI. Besides natural caves, the Mela Trust has constructed a tin shade for the pilgrims. At the time of the fair at Nag Panchami the Trust provides food to the pilgrims. After CHINTAMANI, one goes to CHITRASHALA; a mushroom like large rock formation having many rock paintings. Besides enchanting Nature and rock paintings, the popular belief is that one who visits these places gets ones wishes fulfilled. The distance is nearly 15 km from Kajri and one have to walk on foot hence a guide and physical strength is necessary.

Apart from the above interesting sites of Pachmarhi hills, there are several extra-ordinary places of tourists interest which are very typical to contact till now. Most of the places are not reachable easily. Undoubtedly it may be said that Pachmarhi is a eden of bliss and views.

CAUSES OF DEVELOPMENT OF TOURISM INDUSTRY

Pachmarhi is thought to be as an eden of bliss, natural beauty and magnificent views of all kinds. It is surrounded by high hills, fairly dense mixed forests, showing teak trees, rich fauna, valleys, ravines waterfalls, pools etc. which attract the peoples in all respective wars. It is a hill station most pleasant by its climate. Thus, the main cause of tourism development in Pachmarhi is its natural beauty with pleasant climate.

The peoples of Pachmarhi are very simple and all of them are inhabited by migration. The tribal population is still in poor condition and till now lived in the hilly tract of the area which is far away from the present Pachmarhi town. Some families are found in old Pachmarhi town. The settlement existed generally belongs to military cores or as Government offices while the markets and some hotels are constructed by local inhabited business man in view of personal use along with tourists desire like fooding and lodging. The whole town is engaged in Tourism Industry.

The facilities made after independence are also responsible for the development of tourism industry. Now Pachmarhi has a well developed markets, and all kinds of facilities for fooding, lodging and transportation. The main interesting sites are also linked with metalled roads. The typical places are approachable with footpaths.

The lodging facilities are available in the following hotels (Table 5.1).

Table 5.1 List of Hotels of Pachmarhi

Sl. No.	*Name of the Hotels*	*Phone No.*	*No. of Rooms*	*Location*
A. Hotels Managed by M.P. State Tourism Development Corporation				
1.	Amaltas	2098	04	Near Jai Stambh
2.	40 Avas Griha	2099	40	Near Jai Stambh
3.	Panchvati Huts	2079	05	Near Yuvak Kendra
4.	Panchvati Cottage	2097	05	Near Yuvak Kendra
5.	Satpura Retreat	2097	05	Back to the Forest Deptt. Office on Mahadeo Road
6.	Satpura Retreat (A. C.)	2097	01	Back to the Forest Deptt. Office on Mahadeo Road
7.	Rock and Mainor	2079	N.A.	Near T.V. Tower
8.	Sahkar Bunglow	–	N. A.	–
9.	Tourist Motel	–	–	At Pipariya
10.	Holiday Homes	2099	-	Pachmarhi-Pipariya Road

(Contd....)

B.	**Hotels Managed by Special Area Development Authority (SADA)**			
1.	New Hotel	2017	42	Near P.T.S. Petrol Pump
	Deluxe		06	
	Family Rooms (4 Beds)		08	
2.	Vansthali Cottage	2129	07	Ravishanker Bhawan Road
3.	Nandan Van Cottage	2018	12	Near Jai Stambh
4.	Neelambar Deluxe Cottage	2039	06	Near Jai Stambh
5.	Nandan Van Dormetry	2018	50 beds	Near Jai Stambh
6.	Dharamshala	–	–	Near Bus Stand
C.	**Private Hotels**			
1.	Khushal Kuteer	2136		
	a. (A.C.)		04	Patel Marg
	b. Deluxe		03	Patel Marg
	c. Ordinary		10	Patel Marg
2.	Meghdoot Hotel	2060	08	Bus Stand
3.	Panchwati Hotel	2105	08	Patel Marg
4.	Manjushree	2116	16	Patel Marg
5.	Giri Shringar	2089	16	Near Police Chauki
6.	Hotel Natraj (Deluxe)	2151	16	Bus Stand
7.	Hotel Pachmarhi	2077	37	Patel Marg
	Dormetry	"	50 beds	Patel Marg
8.	Nikita	–	08	Gandhi Chawk
9.	Saket Lodge	2165	07	Patel Marg
10.	Abhilasha	2203	11	Jawahar Chawk
11.	Dipti Lodge	2072	08	Arwind Marg
12.	Panchmahal	2119	07	Arwind Marg
13.	Chunmun Cottage	2133	05	Near Pachmarhi Club
14.	Hotel Shiva	2062	11	–
15.	Hotel Utkarsh	2162	05	Subash Marg
16.	Hotel Sapna	2209	07	Bus Stand
17.	Satpura Safari	2188	N.A.	Patel Marg
18.	Mount View	2062	N.A.	Arwind Marg
19.	Hotel Sukh Nivas	2120	N.A.	Subash Marg
20.	Hotel Mejwan	–	–	Jawahar Chawk
21.	Hotel Banjara	2265	–	Arwind Marg
22.	Khanna Cottage	2156	–	Jawahar Chawk
23.	Hotel Sidhyartha	2126	–	Arwind Marg
24.	Hotel Shilpa	–	–	Gandhi Ganj
25.	Hill View	–	–	Bus Stand

The area provides various types of restaurant for tourists purpose. They provide vegetarian and non-vegetarian food. The details of restaurant are given in Table 5.2.

Table 5.2 List of Restaurant in Pachmarhi

Sl. No.	*Vegetarian*		*Vegetarian/Non-Vegetarian*	
	Name of Restaurant	*Location*	*Name of Restaurant*	*Location*
1.	Prakash Vishranti Grih	Gandhi Chawk	Khalsa Hotel	Patel Marg
2.	Mrig Naini	Gandhi Chawk	Hotel Panchali	Patel Marg
3.	Agrawal Hotel	Gandhi Chawk	Hill King	Patel Marg
4.	Gupta Hotel	Gandhi Chawk	Bagicha Restaurant	Patel Marg
5.	Hotel Akansha	Gandhi Chawk	Swad Restaurant	Saket Hotel
6.	Rahul Restaurant	Bus Stand	Apsara Hotel	Pankaj Talkies
7.	Maharastra Hotel	Arwind Marg	Bombay Hotel	Gandhi Chawk
8.	Surabhi Restaurant	Pachmarhi Hotel	Mayur Hotel	Bus Stand
9.	Hotel Samrat	Bus Stand	Mahfil Hotel	Bus Stand
10.	Chaupati	Bus Stand		
11.	Arpan	Jawahar Chawk		

Restaurant providing regular breakfast like South Indian restaurant (Bus stand), Gupta Sweets (Jawahar Chawk), Lakshman Hotel (Jawahar Chawk) Jain Hotel (Gandhi Chawk), Annapurna (Subash Road), Satpura Hotel (Governments Garden), Palash Hotel (Bus Stand) and Yadav Hotel (Bus Stand) are very much popular in the area. Mobile shops of breakfast are very common at every sites. Drinking water facilities, biscuits, Niboo rus are also available everywhere.

For the purpose of the tourists, there are so many registered guides. Mani Ram Chaudhari (Botanical Guide), Kailash Chandra Parmal, Kishan Lal Chaudhari, Salim Khan, Ramesh Vairagi and Sanjai Ahirwar are the experienced guides of Pachmarhi Darshan. It is found that the matured school boys and girls of different families also works as a guide in their holidays. They enjoy with the tourists and collect the pocket money.

The tourism industry has developed the hotel industry in the town. Most of the people have settled the front of their residence as a hotel and a small cabin of selling the goods for the use of the tourists. The restaurants and hotels where breakfast and meal are provided for the tourists are generally managed by the family members of the manager or 'Hotel Malik'. Servants only serve the materials. Thus the family as a whole enjoys the business. The taxies for approaching the sites may be obtained at the Bus Stand. One can also take cycles and motorcycles by paying reasonable cost per hours. Nothing is impossible for the tourists if he goes to Pachmarhi except his physical labour expended to reach the typical places where any kind of vehicles are not possible to trek on. Thus, the natural attraction and cultural facilities along with the fairdealing of the local population have caused gradually the highness of tourism development.

HUMAN ATTITUDE AND ROLE OF TERRAIN IN TOURISM DEVELOPMENT

As far as the human attitude is concerned, the peoples locally inhabited in Pachmarhi region, which are mostly tribes, never like to develop the tourism industry in the area. They like to pass their life peacefully and alone in the heartland of the plateau. It was the beauty of local terrain and nice climate which have attracted the peoples to come here for minimizing the mental agony. This is the reason why the Britishers came here first. They did not allowed the civilians to encroach the beauty of Pachmarhi. Although the complexity of terrain is not suitable for easy life at Pachmarhi because it is an interior place difficult to reach and manage the daily requirements for the living peoples till now. Pachmarhi-Pipariya is the single route of up and down journey. The moto of the daily coming and permanently living population in Pachmarhi is to see the natural beauty and enjoy with nature and to earn the money from Pachmarhi lovers. Peoples as a tourists go to Pachmarhi for collecting mental peace throwing money for their enjoyment in a limited period of time. Tourists do not care about the degradation of nature. They deeply use the glamour of the nature and enjoy with nature left behind the puzzle of the time. Thus, the tourists arrival causes the development of all kinds of facilities. Tourists need all kinds of adventure and so the local business-minded peoples arrange the attractive materials on a high rates. Hotel industry is also depended on the tourists arrival. It is found by the extensive survey of the area that the rate of the Hotels changes on the basis of the crowdy and off seasons.

The hilly terrain of the region is ornamented with the assemblage of landforms. High hills, plateau, escarpment, mesas, buttes, sleep valleys, khuds, caves, waterfalls, rapids springs, seepages, gorges, ravinousland, multicolour rocks, rock shelters and rock paintings and so many landforms fascinate the Pachmarhi plateau and attract the attention of the tourists. These hills of scenic beauty are thickly blanketed by various important trees mainly teak forest. Rivers and Nalas also fortify the natural beauty along with several water pools available for bathing and swimming purposes which may be possible in nature, is completely reserved in Pachmarhi ecosystem. If it is asked what is life reaching at Pachmarhi, then it may be said that life is a challenge and Pachmarhi provides a chance to meet it. If life is thought to be as a sorrow, puzzle duty, game, song, promise, love and beauty, the nature of Pachmarhi provides a knowledge to solve it, inspires some lessons to perform it, gives energy to play it, feeds feelings to sing it, indicates the words to fulfil it, creates a glamour to enjoy it and fascinates the grace to praise it respectively.

FREQUENCY OF TOURIST'S ARRIVAL

India has an ancient tradition of tourism. It has existed an industry in the informal sector since ancient times and was indulged in by all classes of people. No doubt infrastructure has to be augmented on a much larger scale than higher to but the real thrust to tourism will come only through human resource development as skilled persons are needed at all levels. Also a tourist must be treated as 'ATITHI DEVA' in our traditional hospitable way (Seth, 1996).

Tourism is a quest to discover the wonder of a heritage: Our heritage and those of others with whom we share this planet (Hugh and Gantzer, 1996). The important task is to get tourism accepted as an instrument of development and national integration. Creating awareness about tourism at home and abroad and developing the requisite infrastructure should be our strategy for attaining the desired results (Rehman, 1996). Tourism is as much part of socio-economic development as any other related activity. One can cite any number of examples where the tourism sector has advanced as a consequence of and a concomitant to general development. The industrialised West underline this linkage between tourism and growth. In India, it must be noted that organised tourism began in the 50s with the genesis of planned development. It is another matter that after fortynine years we still have only a minuscule share—0.4 per cent of the world tourist trade which has now been recognised both by the International Monetary Fund and the World Bank as number one generator of income and employment. Apart from being a massive foreign exchange earner, tourism is particularly important to a vast and variegated country like India as it contributes to national integration, promotes social and cultural ambience and plays a key role in socio-economic development. Hopefully this vital sector will receive due attention from those who matter (Seth, 1996).

Domestic tourism in India is quite an ancient concept though the objective was more spiritual than mandane pursuits which prompt the bulk of modern travellers. A great number of 'SERAIS' and 'DHARAMSHALAS' which still dot different parts of the country dating back to several centuries testify to the continuity in cross country movement. Any devout Hindu fells the life's mission is incomplete without a visit to Pilgrim centres and shrines. The Muslims also go to various 'DURGAHS' of Saints they revere. Similarly Jains,

Buddhists, Sikhs and Christians have all their own spiritual destinations which they would like to go at least once.

While pilgrimage is a legacy left to tradition and reinforced by faith, the development of communications and transport and disposable incomes have infused other interests. People wants to go to different places to give families a holiday and share the experience of seeing great monuments, scenic splendour and the delights of seasons in hills and on beaches. This development is comparatively of recent origin and its potential to local development has come to be recognised by both private entrepreneurs and state agencies.

Tourism broadly defined, comprises activities of persons travelling to and staying in places outside their usual environment for leisure, business and other purposes for not more than one year at a stretch. Increase in prosperity along with modes of faster travel and other infrastructural developments has contributed to growth of tourism at a fast rate and more and more people are having access to paid holidays. The tourism development envisages augmented availability and improved organisation of different components of the tourism infrastructure to meet the requirements of the tourists. Various components of tourism infrastructure have to work in perfect coordination to ensure overall performance and availability of linkages in tourism facilities. The development of tourism facilities like hotels, restaurants, recreational activities, transport, shopping etc. is essential for the promotion of tourism and would be meaningless unless the tourists spots have the basic amenities and infrastructure like water, electricity, roads, telecommunications etc. The tourism planning, therefore, necessarily entails an integrated development of tourism infrastructure and facilities in the area (Makhija, 1996).

The beautiful nature, the exciting hills, the charming river valleys, the imotional rivers and rivulets including waterfalls, rapids and pools, the glittering stones by sun rays, the smiling gardens with flowers, the decorating fauna, the fascinating wildlife and bio-diversity, the attracting rock shelters and rock paintings, the gradually flowing sweet winds and promoting mental peace, grace and glamour through pleasant weather and climate etc. have made the Pachmarhi an enchanting land of amazing grandeur. There is a tremendous scope of tourism and surprisingly there has been enough tourist flow to this region from time to time.

On the bases of the extensive survey of the area and asking questions from the peoples of Pachmarhi it is generally found and most of the respondents have indicated that Pachmarhi receives more than 20 lakhs tourists in a calender year. 'Shivaratri', 'Nag Panchami' and 'Rakshavandhan' are thought to be as the main fairs of Pachmarhi. On these occasions, about 5 lakhs fair-tourists are believed to come in each fair of Pachmarhi. 'Chaitra Nava Durga Pooja', 'Somvati Amavashya' and 'Rohini Sankranti' are the small fairs when about 10 to 20 thousands of peoples in each festival are noted to come here to see the beautiful places and for the holy bathe in rivers and 'Darshah' of 'Lord Shiva'. In the season, generally one thousand peoples and in off seasons about 300 tourists per day come to the place. On festival times the tourists generally come from the different localities of the country including the local peoples of M.P. state. The cultural programmes like exhibition, folk songs, dances, etc. are generally to be organised at the time of 'Pachmarhi Utsav' held in the month of October each year. This time, a large number of tourists come to enjoy the 'Pachmarhi Utsav'.

TOURISM AND POSSIBILITIES OF AREA DEVELOPMENT

Agricultural development, industrial development, and tourism development are three ways to bring money and business into a given region. Tourism development is probably the quickest method of the three and it has become the part and parcel of area development of Pachmarhi. For centuries our people travelled widely to exchange knowledge, to know each other, to buy and sell goods, to attend trade fairs and exhibitions and to do research etc. But in Pachmarhi, the tourists of all the times have come to robber its beauty. Tourism development encompasses practically all fields of economic activity. Apart from generating employment and augumenting national income, tourism offers unlimited opportunities for bringing man and his heritage together (Negi, 1996). As far as the Pachmarhi, it is true that all kinds of development like hoteling, restaurants, transportational activities, markets, etc. have totally caused due to tourism. Tourism has given a new culture of socio-economic development in the area.

Tourism development is a process of conflict and compromise. It is a matter of striking a balance between the competing groups, i.e. between the private interests and the public interest, commonly known as the national interest. It is a means of increasing the level of economic activity of the host nation through sale of products and services into the travellers. Travel to an area provides a basis for developing tourism as an export industry. The amount of tourism expenditure that remains in the area provides a source of income to local residents and is considered as the direct effect of visitors expenditure. The secondary effect comes as the money paid by the visitors to the business is utilised to pay for supplies, wages of workers and other terms used for the benefit of visitors.

Tourism multiplier is estimated on the basis of sales and output employment or payroll or other variables. As the number of visitors making purchases increases, demand for products and services produced in the area is also increased which directly or indirectly creates employment opportunities for the residents. The local inhabitants of Pachmarhi generally engaged himself with family members in any type of business related to tourists. Most of the servants working in restaurants and hotels are settled by the services needed for the tourists. Thus, the tourism development has provided an external source of income to the peoples of Pachmarhi.

The social aspect of tourism development in Pachmarhi is related to the new job opportunities and influx of new income to the area. As such, a broad tourism development programme should incorporate measures for educational and training facilities for the local peoples. The tourism development has caused in the mind of the peoples of the area to act as the guide and thus the local residents, even the students are seen in this business. In one of the research studies, it was revealed that the 'hotels' cultural shock on the community is slowly wearing off and its benefits are outweighing the negative effects. It is found that the peaceful area of Pachmarhi is now giving a sense of western culture and thus the economic benefits slowly create mental pollution.

The tourism development of the area has a powerful answer to some of the environmental problems such as poor water supply, poor electric supply, inadequate sanitation and sewage facilities, deficient nutrition, bad housing conditions, disease and sickness and vulnerability to national disaster. The tourism development has provided some more facilities in the area, for example, the major tourists centres have been linked with metalled roads where taxies are available to contact the places. The natural scenes have been also fashioned by some artificial facilities. Thus, a new tunning is common everywhere. Tourism development revives and rejuvenates the local culture too. Practically all our local artists, craftsmen and those employed in performing arts are employed professionally, which in turn sparks off a renewal of interest of local residents in their own cultural heritage. The business of arts and paintings and camera printings have been most effective in the area. It is also believed that the area is very rich in rock paintings since the ancient times.

Pachmarhi has taken its birth on the basis of tourism development. The natural environment which has been reshaped till now and going to be changed in near future is totally affected by tourism development. The travellers have developed the area since the time of Pandavas to the present state of knowledge.

CONCLUSION

On the basis of the above discussion, it may be said that Pachmarhi is a place of worship, beauty, prayer and pilgrims. Truly speaking, it is an eden of miss, kiss and bliss.

REFERENCES

1. Bansal, S.C. 1991: *Pachmarhi: A Trekker's Paradise,* An introduction to Pachmarhi with all necessary information alongwith beautiful coloured photographs and guide map, Kamal Kishore Dhoot Publishing Company, Pachmarhi.
2. Hugh and Collen Gantzer, 1996: The Tourism Quest, *Yojana,* Ministry of Information and Broadcasting, Publication Division, New Delhi, Vol.40, No.8, August, pp.11-13.
3. Jayaswal, D.P. 1952: *Pachmarhi Darshan* (in Hindi) Kailash Printing Press, Pipariya, Hoshangabad, pp.1-99.
4. Makhija, B.N. 1996: Tourism Planning in India, *Yojana,* Ministry of Information and Broadcasting, Publication Division, New Delhi, Vol.40, No.8, August, pp.29-31.
5. Negi, J. 1996: Tourism and Development: Some issues, *Yojana,* Ministry of Information and Broadcasting, Publication Division, New Delhi, Vol.40, No.8, August, pp.33-37.
6. Rehman, S.S.H. 1996: Strategy for Boosting Tourism, Yojana, Ministry of Information and Broadcasting, Publication Division, New Delhi, Vo1.40, No.8, August, pp.17-18.
7. *Pachmarhi: Pahariyon Ki Hari Mani* (in Hindi) M.P. booklet published by M.P. Tourism Development Corporation.
8. *'Satpura Shringarika' Pachmarhi* (in Hindi) Bharat Photo Studio, Pachmarhi.
9. Seth, R. 1996: Tourism in India: An Overview, *Yojana,* Ministry of Information and Broadcasting, Publication Division, New Delhi, Vol.40, No.8, August, pp.19-21.
10. Seth, M. 1996: Tourism in India: Some Thrust Areas, *Yojana,* Ministry of Information and Broadcasting, Publication Division, New Delhi, Vol.40, No.8, August, pp.7-9.

Impact of Natural Hazards and Human Interferences on Existing Ecosystem

TYPES, TRENDS, DIMENSIONS AND IMPACT OF NATURAL HAZARDS

Everything in environmental history is changing in nature. One of the greatest problems facing mankind today is to achieve a nice balance of nature between a rapidly increasing population and human thrust. Land degradation and loss of fertile top soil have challenged the human race while techno-scientific encroachments have depleted the natural resource base of this planet. Thus, the wholesale cutting and ravaging of our forests, the removal of native sod cover, the senseless overgrazing, unnatural drainage operations and agricultural techniques, draught damage, floods, dust storms etc. are the major problems alarming the need of environmental improvement (Prasad, 1991).

The upland region is highly dissected having incised deep valleys of seasonal torrents. Khohs and khads are the major degraded lands due to fluvial actions. The forest covers are mostly reserved and sparsely populated by tribal population. The sandstone country produces very poor soil for agriculture. Sand accumulation in the downslope areas, rocky wastes, uneven terrain loss of biological diversity etc. are thought to be as the other problems of the area.

Forest as a renewable resource contributes goods (mainly food, fuel and medicine) and services to the people of the region. The forest culture of Pachmarhi is richly illustrated since Mahabharat period. The forest destruction and denudation are principally governed by population encroachment and livestock leading to enhance requirement of food, timber, fuel, wood and grazing respectively. The existing tribal population and local communities have misused the forest resources. The ravine tracts of the region also denote forest degradation because the deepest khohs and khads are the indicator of forest loss. The another cause of deforestation is the construction of hill roads in the area. It is remarkable that the tourists interest sites are now well cemented by metalled roads. The holy places have added many infrastructural facilities like Dharamshalas, recreation sites etc. These activities have finally caused the clearance of forest cover in the region.

The problem of stone quarries are also common. The area is rich in an excellent quality of sandstone. Stone quarries are generally marked along the hill sides for the collection of building materials by tribal population and for the use of building-making in urban areas. The morrum quarries are also seen. It is apparent that these quarries accelerate erosional capacity initiating various forms of mass movement.

Mass movement and land sliding are very common features generally caused by accelerated soil erosion. Mass movement and land sliding are also caused due to steep slopes of the area.

Collapse of rock roofs in limestone area is also noted in most of the caves found in the area. The Jatashanker hill's cave denotes this type of problem at a greater extent.

Vertical erosion, rilling and gullying are also very common along the rectilinear slope profile of the region.

NATURE OF HUMAN ATTITUDE TOWARDS THE ECO-SYSTEM

Since long, Pachmarhi is thought to be as the tourist's place. It is used by the coming generations as the place of providing natural gift, glamour and enjoyment. Thus, the beauty of the place have been stolen by the peoples day by day whenever they approach the nice campus of Pachmarhi. It is praised by all in various respective ways. It is the highness of Pachmarhi's landscapes that they have ever reflected virgin ideas of glamour in their sovereignty. Degradation is a common factor in an every state of landform but it seems that the changing panorama of Pachmarhi due to human interferences is increasing gradually. Natural processes have caused various peaks, deep valleys, springs, waterfalls, rapids, caves, khohs, khads etc. as the assemblage of fluvial landforms and clearly indicate the gradual degradation of the existing landforms but in view of associating groups of local peoples and tourists in Pachmarhi, these erosional landforms are the attractive places fortifying the charms of the area. Thus, the changing attitude of the plateau due to destructive natural processes is the boon for Pachmarhi which is eden of bliss.

In view of tourists, it is a nice place. It is a natural world of dreams. We are dreaming here but leaving here some wastes after rest. It is found that the interesting sites of tourists are now encircled by dirty loose materials. Peoples never follow the rule of making the environment neat and clean. The hotels and restaurants are smiling in western culture. The local culture is also badly affected by the tourists. Although the existing population of Pachmarhi never like for this type of behaviour but they are unfortunately related to tourists arrival directly or indirectly. They love Pachmarhi because tourists love Pachmarhi. Love is very common in both the side. The major factor initiating the natural love is the natural attraction of Pachmarhi. Seeing, the scenic beauty of Pachmarhi, it may be said that we are enjoying with Pachmarhi without any care of present and past and leaving behind sorrow kits. None of them have been seen in a thinking to improve the nice picturesque of Pachmarhi.

HISTORICAL REVIEW ON THE CRUX OF ENVIRONMENTAL DEGRADATION

According to the Untivatika copper plate inscription (*Hoshangabad District Gazetteer*, Vol. VIII, 1908) the eastern part specially around Sohagpur, was the territory of a king Abhimanyu of the Rashtrakut clan of a place, Manpur, in about the 7th century. The Bhopal inscription of Udaya Varma mentions that the grant of Gunaura village in the Narmadapur PRATIJARANAK in the Vindhya mandals in A.D. 1199 (*Hoshangabad District Gazetteer,* Vol. XVI, 1927). This and the Harsud inspection of Delpaldeva (A.D. 1208) suggest the domination of Paramara kings of Dhar (Smith, 1928, 57, 58) in the central and western parts of Hoshangabad.

In the mediaeval period this tracts appears first in the historical records during the reign of Sultan Hoshangabad Ghori of Malwa (A.D. 1405 onwards) who built a small fort at Hoshangabad, along with two others at Handia and Joga. In his expeditions against the king of Kherla near Betul, he always took the route through Harda and Hoshangabad. After the fall of Mandu in 1567, Malwa was annexed as a SUBAH of Mughal empire. Handia, 21 km north of Harda was the seat of a SIRKAR with its MAHALS at Seoni Malwa, Harda and Bichhola to the south of the Narmada. However the fort of Ginnurgarh, northwest of Hoshangabad proper across the river Narmada continued to be under the Gond kingdom of Garha-Mandla.

In the early 18th century, the area of Hoshangabad district was divided into seven political divisions. The northern parts of Seoni and Harda tahsils were under the Mohammadan FAUJDAR of Handia, while the Rajwara PARGANA in the eastern part was held by four RAJAS feudatories to the Gond kingdom of Garha-Mandla. Even Hoshangabad was then subject to the control of Ginnore.

In A.D. 1722, the territories of Ginnurgarh including the fort of Hoshangabad felt to Dost Mohammad Khan, the Nawab of Islamnagar and founder of Bhopal dynasty. The Peshwa Balaji Bajirao captured the Handia SIRKAR, west of the Ganjal river in A.D. 1742 and replaced the Mohammadan governor there with his own AMILS, who later in A.D. 1750 also forced the Raja of Kalibhit (Makrai) to cede half the number of his villages to the Peshwa by a treaty in

1750. By the close of the 18th century the tract was made over to Sindhia. The remaining principalities falling in Seoni-Malwa, Hoshangabad and Sohagpur tahsils, east of Ganjals gradually came under the occupation of the Bhonsla Raja of Nagpur between A.D. 1740 and 1775. Beni Singh his SUBAHDAR at Bhanwargarh also captured Hoshangabad fort in 1796. From 1802 to 1808, Hoshangabad and Seoni were lost to Bhopal Nawab but were ultimately regained by the Bhonsla Raja of Nagpur in 1808.

In the last Anglo-Maratha war of 1817, Hoshangabad was occupied by the British and was held under the provisional agreement made by Appa Sahib Bhonsla for its cession in 1818. In 1820 district ceded by the Bhonsla and the Peshwa were consolidated under the little 'Saugor and Nerbudda Territories and place under an agent to the Governor General, residing at Jabalpur, Hoshangabad district at this time included the territory from Sohagpur to the Ganjal river. Harda and Handia remaining with Sindhia. From 1835 to 1842 the areas of Hoshangabad, Betul and Narsimhapur districts were kept amalgamated into one with headquarters at Hoshangabad. As a consequence to the Bundela rising of 1842, they were again separated into three districts as before and the designation of the officers holding general charge was changed to Deputy Commissioner from that of Assistant to the Agent. In 1844, the Harda-Handia tract was also made over to the British by the Sindhia as part of the territory assigned for the support of Gwalior in 1860. Nemarwar Pargana north of the Narmada which was also made over in 1844 was returned to Sindhia in 1860. The Kalibhit tract of Harda tahsil was transferred to Harsud tahsil of East Nimar in 1905.

Prior to the 30 years settlement of Charles Elliott (1865) the district was divided for administrative purposes into 42 Talukas. To avoid such minute subordinate sub-divisions, the talukas were distributed at that settlement into six Parganas which going from east to west, were named Rajwara, Sohagpur, Hoshangabad, Seoni, Harda and Charwa. This however was soon superseded by the present division of four tahsils viz. Sohagpur, Hoshangabad, Seoni and Harda. In about 1951, Pachmarhi formed a tahsil but was reduced to a sub-tahsil as before with an additional tahsildar posted their subsequently.

Historical evidences denote the facts that Narmada attract the primitive men to establish himself near its edges as early as in the early palaeolithic and microlithic ages, when mammals roamed about freely and the earth was passing through its early Pleistocene period. In recent excavations ample material in situ has been picked up from the soil of the area to prove this. During 1958-59 and 1963-64 when explorations were conducted in Hoshangabad and in the upper Narmada basin from Amarkantak to Harda, 35 fossilferous and implementiferous sites, some of them with instructive sections and extensive deposits of cemented gravel were discovered. From these sites more than 1000 palaeoliths including hippopolamus and other unidentified animals were recovered. It is however noteworthy that there is nothing like a stratigraphic evolution in the Narmada Palaeolithic industry. In a gravel bed at Pipariya, forty artefacts representing the earliest phase of the chelleanage and illustrating the evolution of hand axe from the pebble to the earliest Abbevelian stages, were found. They were huge heavy and crude (*Indian Archaeolgy,* 1959-60).

From the oldest period till 1863, the area was well known for war and peace but Pachmarhi used as a shelter of peace during the warfare—because the area shows various rock shelters. The rock shelters of Pachmarhi and Adamgarh hills have preserved interesting archaic paintings. These paintings have revealed various interesting faces of primitive life in this area. The most important painted shelter is Bhim Baithak. The subject matter of these paintings is mostly hunting of animals and birds with heavy sticks, lows and arrows. Scenes of domestic life have been found particularly in the rock paintings of Pachmarhi. It is interesting to note that a large number of microliths including blades, points, triangles, lunates, crescents, trapezes about 40 per cent of which showed geometric shapes were collected from the eighteen trenches that were laid in the rock shelters and their close vicinity at Adamgarh. A few flake-blades and fossil remains were also collected from Sandia and Seoni-Malwa.

According to a local myth Sohagpur was known in ancient times as Sonitpura. As per the narration of Harivamsa. Sonitpura was the capital of the Demon king Banasura whose daughter Usha was abducted by Krishna's grandson Aniruddha from the capital city (Dey, 1927). It is also believed that the five Pandava brothers spent a major part of their life in exile in the Hoshangabad district. To substantiate their imagination the local people show the five rock caves of Pachmarhi as the abode of the Pandavas and a spot on the Sandiaghat, where they once took rest and cooked their food. Similarly they associate Siva with the Mahadeo hills. The 'REVA KHANDA' of the 'SKANDAPURANA' contains a vivid account of the Narmada and the sacred 'Tirtha' on its banks.

The Mugal history recites the fact that during the Shahjahan's and Aurangzeb's reign the Province of Malwa continued to enjoy great importance due to its central position and richness in important agricultural products. Important routes connecting Northern India with the Deccan also passed through this region (Sarkar, 1912 and 1925). Sinha (1994) advocates that the Sohagpur Pargana suffered most from the devastation of the Pindaris during British period. It was so entirely deserted between 1803 and 1818 that not more than three or four villages in the Babai tract were left inhabited and in the vast majority of cases the Patels and cultivators had strongly disappeared. Besides, there was left not a single village in the Hoshangabad district which was not burnt once or twice in the course of these fifteen years. The people used to crowd themselves under the protection of such forts as those of Sohagpur, Seoni Malwa and Timarni for their mutual defence.

It is also reported in the early history that Appa Saheb was wandering in the hilly tract and inaccessible forest region of the Hoshangabad district. During his flight through the Mahadeo hills he took shelter at Pachmarhi. From there he issued 'SANADS' in August 1818 to Sahib Singh, Raja Singh and Bal Singh for the tract of Sohagpur, Barha and Bachai, respectively and utilized their services for raising troops for his service (Report 1939). Besides receiving help from a number of Gond chiefs he befriended the Pindari leader Chitu also. Gradually, Appa Saheb collected a force of 2000 troops consisting of the Arabs and about 2000 Gonds. Mohan Singh, the Jagirdar of Pachmarhi, and Moti Bai and Rangoojee, the Gond chiefs became his close adherents.

Finding it difficult to arrest Appa Saheb the British proclaimed a reward of one lakh rupees for his seizure (Adventure of Appa Saheb, 1939) but he could not be captured. Appa Saheb made good his escape from Pachmarhi on the night of the 1st February 1819 accompanied by Chitu and Sheikh Daula, the Pindari chiefs. It is also reported that Tantya Tope also continued his march almost straight to south by the way of Pachmarhi hills to Jamai, a police post 26 miles from Chhindwara (Dharam Pal, 1955). In 1859 Nawab Adil Mohammad Khan of Ambapani entered in the Hoshangabad district (Chaudhary, 1957). He was joined by Babhut Singh, Jagirdar of Harrakot in Mahadeo hills. Adil Mohammad, however, retired soon. But Babhut Singh carried on depredations, aided by the Fatehpur and Sohagpur Rajas. A detachment of Madras infantry and military police marched against him in August 1859. A fight took place near Denwa river with Babhut Singh, who was repulsed with some loss (Chaudhary, 1957). At length Babhut Singh and his chief subordinate Holibhai were caught in Jaunary 1860.

After the overall discussion on the history of the region, it maybe said that the area of Pachmarhi is mostly utilized as the place of shelter since long to the present time. It was the peaceful zone occupying hills and thick forest cover. Previously it has been used as a shelter and now it has so nice balance of ecosystem to attract the tourists for their recreation. It seems that Pachmarhi has no loss but gradually has improved the number of their lovers who enjoy with the heart and soul of the virgin land without any care and hesitation. It is the only drawback for the beauty of Pachmarhi.

CAUSATIVE FACTORS OF ENVIRONMENTAL DEGRADATION

It has been marked in previous discussion that Pachmarhi hills are in degrading condition due to the two major factors — (i) The natural processes involve in the degradation of natural beauty of the region, and (ii) the human interferences release time to time in the region. The environmental degradation due to natural processes or geological processes is universal in nature because the upliftment and peneplation of the terrain are the initial and last stages of the cycle of landform origin. Pachmarhi hills are thought to be in late mature stage. It is gradually decreasing downward due to retreating and declining of the slopes. The degradation has been enlarged due to the tourists encroachment. But, in all enquiries, it is not in serious dimension. A slight jerk has been observed by human carelessness with nature.

GOVERNMENT AND PUBLIC ISSUES TOWARDS ENVIRONMENTAL DEGRADATION AND MAINTENANCE

Pachmarhi hill resort is a unique hill resort which has so many things that one can not find in any other hill stations. Though this hill resort is being developed by the government but the pace is very slow and therefore you might have lacked modern facilities that you could get at other hill stations, but very candidly spoken it can be said that the uniqueness about Pachmarhi lies in its naturalness and undiluted environment. You must have seen that modernities have inferred very little with the natural spots and Nature of Pachmarhi is in its most natural form (Bansal, 1991). Hotels, restaurants, hospitals etc. and number of other facilities are easily available in Government and public sectors of this area. The facilities provided by public sectors are not advance in nature. It is observed that the growth of town and sewage of the settlement cause too much pollution as it is seen near Ganesh hills. The Nala before the military campus is also very dirty place which need the proper management. Most of the tourist sites are very alone and almost dirty. No precautions have been taken to remove the waste materials. The picnic spots, gardens and boating sites need further improvement. The facilities are very few in all the places. The businessmen have no interest to develop the interesting sites, therefore, the challenging efforts will be made by the government at every spot of the region. Even then the most interesting sites like Jatashanker, Gupta Mahadeo, Bara Mahadeo, Dhupgarh, Chauragarh and the sites of various interesting watersfalls have not a proper facilities to reach and enjoy there. The nature of Pachmarhi has self decorated her beauty and charm. It is not a gift of local peoples or through government efforts.

CONCLUSION

Apart from the above discussion, it may be concluded that Pachmarhi is a best hill resort embracing her natural beauty. It is still neat and clean after roughly using by tourists. The sign of pollution which is tolerable, has been only marked in urban area but the beauty of Pachmarhi need some artificial arrangements which is only possible by governmental efforts. It is the best hill station of Madhya Pradesh and the queen of Satpura.

REFERENCES

1. *Adventure of Appa Saheb*, Nagpur, 1939.
2. Bansal, S.C. 1991: *Pachmarhi; A Trekker's Paradise,* Kamal Kishore Dhool Publishers, Pachmarhi.
3. Chaudhary, S.B. 1957: *Civil Rebellion in Indian Mutinies,* 1857-1859, Calcutta.
4. Dey, N.L. 1927: *The Geographical Dictionary of Ancient and Mediaeval India,* London.

5. Dharampal, 1955: *Tatya Tope,* New Delhi.

6. *Hoshangabad District Gazetteer,* 1908, Vol.A.

7. *Hoshangabad District Gazetteer,* 1927, Vol.A.

8. *Indian Archaeology: A Review,* 1959-60.

9. Prasad. G. 1991: Morphological Processes, Landforms and their Significance in Environmental Management of Chitrakut and its Adjoining Region, Ph.D. thesis (unpublished) submitted to Kanpur University, Kanpur.

10. Sarkar, J.N. 1912 and 1925: *History of Auranzeb,* File 2 and 5, Calcutta.

11. Sinha, A.M. 1994: *District Gazetteer, Hoshangabad,* Bhopal.

12. Smith, V.A. 1958: *Akbar, the Great Mughal,* New Delhi.

13. Smith, V.A. 1928 and 1957: *The Cambridge History of India,* File 3 and 4, Cambridge.

Environmental Management

INTRODUCTION

Environment has been defined as the sum total of all conditions and influencess that affect the development and life of organisms. The system preserves both physical and cultural attributes as a boon as well as a curse for existing civilization. The so-called danger of environmental deterioration caused by the interactions between physico-cultural attributes has attained serious dimensions at places. Its study relates to environmental toxicology which in turn needs an urgent attention of eco-management in all possible techno-scientific ways (Prasad, 1991).

Planning is considered to be a process of development through the exploitation and utilisation of all types of resources whether natural or human while environmental planning implies the optimal utilization of the earth resources, both renewable and non-renewable for development activities conservation of what is rare and precious in nature and preservation of the quality of environment for the healthy growth of life (Singh, 1983). Management implies a conscious choice from a variety of alternative proposals and further more that such a choice involves purposeful commitment to recognized and desired objectives. Environmental management is difficult to define because even the term environment in itself is complex as it is understood differently by different sections of society. The objectives of environmental management are complex, varied and even conflicting, and the alternative strategies are divergent (Singh, 1991). The concept of environmental management is generally related with the environmental model which assures the basic necessities with their important limits and includes the challenges to be faced and the policies to overcome the problem (Prasad, 1991).

Human activities of modern 'economic and technological man' have disturbed the harmonious relationships between the environment and man. Thus, environmental management is the process to improve the relationship between man and environment so that the quality of both the environment and human society, may be improved. This improvement of relationships between man and environment may be achieved through check on destructive activities of man, conservation, protection, regulation and regeneration of nature. Environmental management, thus is related to the rational adjustment of man with nature involving judicious exploitation and utilization of natural resources without disturbing the ecological balance and ecosystem equilibrium (Singh, 1991). Environmental planning and management is therefore compromise between ecosystem and ecological balance and human material progress and thus environmental management must take into consideration the ecological principles and socio-economic needs of the society.

As a matter of fact, environmental management involves the protection of environment in all possible ways, enhancement of economic values of the environment and its resources and preservation of the environment for future generations. Environmental planning and management include the whole biological world. There is not any artificial boundary at regional, national or global level. But the processes of management and planning differ in different regions by changing human interests. The question of eco-management issues are indeed ever-widening. The purpose of eco-improvement is to develop it as a serious subject with a positive approach enabling us to wipe out the huge ecological deficit on the one side and prevent future environmental damage by making sustainable development on the other.

The aim of eco-management planning is to identify the problems of problem areas and the way of corrective actions which may lead sustainability in physico-cultural development. Eco-management issues in India were raised by Pandit Jawahar Lal Nehru as early as in 1957 when the question of environment in India was not even talked about. He wrote "we have many large river valley projects which are carefully worked out by engineers. I wonder, however, how much thought is given before the project is launched to having an ecological survey of the area and to find out what the effect would be to the drainage system or to the flora and fauna of that area. It would be desirable to have such an ecological survey of these areas before the project is launched and thus avoid an embalance of nature. There are so many organisations, programmes and projects which are actively engaged in the study of man-environment relationships, the effects emanating from such interactions and their possible remedial measures at international levels. It is a healthy sign that International cooperation is available for the mitigation of severe environmental problems affecting the mankind at global level."

REVIEW OF MANAGEMENT PLANS

Pachmarhi hills need the variety of management plans. The existing eco-system exhibits the vivid picture of terrain characteristics imbracing some serious impact of natural and human based eco-problems.

Environmental crisis is a common question appearing at every step, but it becomes more serious if it is growing by nature itself. The problems caused by ravination and the badly encroachment of khohs and khads in the area have been treated as natural problems. Thus, the conservation planning of natural resources and the improvement of the terrain is the first priority in maintaining the ecological balance of the region.

Man's interferences in natural phenomenon have superimposed a new assemblage of terrain characteristics. He has used and reused the nature without much care for future generation and has changed the landscape according to his choice and needs. Obviously, the man-made crisis has precipitated more due to his involvement, positive or negative and unchecked interactions with nature. Thus, the planning should be implemented at the level upon which man has engaged himself to consume the nature.

The nature becomes more poisonous, if the physio-cultural ecosystem together display the negative role. It is obvious that most of the serious eco-calamities are set in hazardous conditions due to ecological imbalance in nature For example, the soil erosion is caused by the running water on the one hand and is further accelerated by faulty management of terrain, deforestation, stone quarrying etc. on the other.

Physical environment has been dealt rather harshly by man through his advertent/inadvertant faulty actions. So, we must recognize the man, along with other life forms, we have always been subjected to environmental stresses over which he has no control. But protective and evasive action is possible where advance sophisticated knowledge and methodology are available. We shall be foolish, if we don't take the opportunity to arm ourselves with information about these forms of environmental hazards. Where do such things occur? When can they be expected? Can warning system be made effective? What should be done when disaster threatens? Broadly speaking such information comes under the process of environmental protection. Although we have not given a full account of protection measures, we will explain how and where these dangerous natural phenomena operate the system and how they can encountered?

The nature and magnitude of eco-management planning have been suggested differently in various parts of the world depending upon the nature of the problems and the technologies available in the region. In the study region, the problems have been tackled in a superficial way because we do not have the requisite infrastructure to monitor the ecological changes and the possible hazards which threaten our life.

On the basis of the personal experience and some dialogue with the persons concerned in the environmental planning, the researcher has suggested a list of plans related to topo-management, climo-management, Biome management, watershed management, Groundwater management, techno-management, tourism management etc. All these plans are necessary for the eco-decoration of the area.

SUGGESTION FOR TOPO-MANAGEMENT

Pachmarhi is known as 'Eden of Bliss' due to his topo-culture. It includes in her lap, the nice assemblage of landforms. Terrain topography is caused by fluvial dominance of the area. Landform complexity is the major characteristics of the region. Moderate to high stream frequency, drainage density moderate to very fine drainage texture, moderate to steep average slope, moderate to high relative and absolute reliefs and moderate to high drainage dissection have ornamented in the area, a unique type of topo-features. Springs, seepages, rapids, waterfalls, caves, khohs, khads, ravines, deep valleys, gorges, mesas, buttes, escarpments, ponds etc. are so many landforms attract the lovers of Pachmarhi. The hills are dissected and have a steep slopes. From foothills to hillstop, the terrain is thickly dotted with dense forest cover. Most of the river draining over the hill country make the sinuous path with narrow deep passage. In the upper part of the river, stony bed occupying big boulders and thus creates several breaks in slope. Rolling surfaces are very common over the plateau top.

Land degradation which is expressed as soil erosion or ravination is well-known phenomenon and has become a much more comprehensive subject of environmental crisis of the modern world (Eckholm, 1976). It has the most serious impact in the poorest countries of the world and on the poorest peoples (Prasad, 1985 and 87). It means primarily the lack of food for the growing populations of most countries and this is a major and ever increasing global problems (Vink, 1983). The management of soil erosion to save the topo-prosperity of the region, the following measures may be suggested:

(a) Ravination can best be prevented and controlled by keeping the affected ground covered with vegetation because the maintenance of an adequate vegetative cover is the surest way to control the burst of heavy torrential rain which in turn causes environmental crisis.

(b) Gullies in the early stage of their development can be effectively dealt with ploughing and seeding with grasses. Once the grasses grow and a sod cover is formed further erosion may be prevented.

(c) Where gullies are too deeply entrenched and it is difficult to plough the land, it is necessary to construct retaining walls perhaps of brush wood or wire netting with a straw gaurd to catch and hold the eroded soil. Great charms may demand the construction of concrete or stone dams.

(d) Terracing is important in sloping land under certain condition. But, if used improperly, it can be ineffective and may be harmful.

(e) The sloping country which is prone to erosion and cultivated should be practised by contour ploughing because ploughing up down slope will clearly assist run off and predispose towards gully formation.

(f) Strip cropping is a suitable measures in most of the cultivated areas. Strips of grass and soil binding are of ravine reclamation plans and may be discussed in the following manner:

  (i) The ravines themselves can be brought under cultivation where these are sufficiently wide by the

formation of a series of terraces one upon another along the slope. Where ravines are moderately wide they can be terraced and must be used for growing horticultural crops.

(ii) The serious regime of ravinous belt must be converted into a pond. The pending condition must be surrounded by trees.

(iii) The small and medium gullied land surrounding the towns should be used for settling landless labourers, lower class peoples and workers and as well as for playground.

(iv) The khohs and khads must be controlled with the construction of check dams or gully plugs in the ravines at proper intervals.

(v) Crop management practices including the use of a legume or grass crop in rotation should be adopted in an integrated watershed planning of ravine lands. Cover crop should be harvested when the main crop is harvested. Continuous farming by planting rows of crops and grasses and terracing may be followed in sloping areas.

The topo-management is still in progress only in urban area of Pachmarhi. The most of the hill sides are still in natural form. They are out of planning process because of lack of planning facilities.

CLIMO-MANAGEMENT

Climate is thought to be as the universal factor controlling the entire eco-system of an area. Pachmarhi experiences pleasant climate and attracts the easy growth of various biological elements. It is the climate which has fashioned the biological diversity in the region. Various species of flora and families of fauna are seen due to positive response of the climatic conditions. It will be challenging efforts to manage the climatic regime. In view of the suggestion, it may be said that the plateau region must be ornamented through the new plantation of important trees and herbs and by the construction of some ponds and lakes over the plateau so that the climate of the area which is pleasant in nature may stand still in pleasant manner. Heating and cooling of the atmosphere have been also maintained with the above processes. The concept of urban and industrial growth must be minimised in this eco-sector for balancing the climatic factors.

BIOME MANAGEMENT

Pachmarhi hills is too much rich in flora and fauna. The bulk of the forests in the area fall under tropical dry mixed deciduous, tropical moist mixed deciduous, tropical semi-ever green and sub-tropical wet types based on climatic and ground conditions. Some rare species of plants and herbs are also visible here. The region shows 770 vegetation types of 452 plant species under 101 families. The natural vegetation denotes 247 types of trees and 531 types of small leaf plants. Owing to the plentiful supplies of food and water throughout the greater parts of the forests, the area used to be an ideal abode for wildlife. But during the past 50 years the poaching has badly affected the animal population. The maximum damage was done by the persons having crop protection licences. Similarly intensive working of the forests has caused disturbance in the forests which has affected in breeding of the hervivorus animals. The reduction in their number has also in turn affected the number of carnivorous animals. The carnivore has gone down terribly. The tigers and the panthers which could be seen with little effort in forests are now rare in occurrence.

Deforestation is the primary cause of land degradation in Pachmarhi region. Deforestation is caused due to unchecked cutting of the trees by the contractors, overgrazing, stone excavation and morrum quarrying, supply of fuel, fodder and timbers for local peoples. The major part of Pachmarhi protected forest, the southeast part of the area and the scattered patches of the forest along the Denwa river in the north have been affected by deforestation. Illegal clearing of the forest by local people is also most common in thickly populated areas of the region. The tribal population generally supplies the fuelwood to fulfil their primary needs to the local population.

The following suggestions may be proposed for the biome management:

(a) Most of the tribal populations are intimately tied to the forests as they depend for their entire needs upon the forest resources—fuel, fodder and small timbers. Thus, the plant species providing fuel, fodder and timber must be planted along the sloping marginal land of the hills where the population of local people are most common.

(b) The indigenous and exotic fast growing plant species are thought to be useful in plantation along the well drained sloping ground, ravination areas and the lower section of the residual hills.

(c) The agricultural areas of plateau top and the faces of the scarps should be harvested in a very careful and controlled manner as to avoide soil erosion and tree damage. The plantation of rooting plants are the best solutions in stony bed areas.

(d) The areas of the concave hill slopes, river sides and undulating country must be managed by planting suitable plants like tall trees in concavity areas, smaller trees and bushes in the undulating river sides and small shrubs and herbs in the level ground or the marginal flat lands. This will enable the plants to make full use of sunlight, water and soil resources.

(e) The vacant areas of the villages and agricultural lands and the lands along the roads can be utilized for commercial or exploitative forestry.

(f) Fuel, fodder and timber plants should be grown in the ravinous belt.

(g) It would be better if the areas near the Pandav caves and Jatashanker hills be converted into 'Van Vihar', the

botanical garden and as a forest research centres. The forest department should take up the job and establish a research centre to conserve the flora and fauna and also to provide attraction to the tourists. The proposed research centre will experiment and suggest precisely the nature of plants to be grown at the specific spots.

(h) A national park and sanctuary may be developed near Gupt-Mahadeo and Dhupgarh hills areas.

In addition to above mentioned suggestions, the conservation of biotic ecosystem including the preservation of plant species wildlife and living resources available for the tribal population is of vital importance. The medicinal plants and herbs must be conserved and recycled seriously and scientifically. Wildlife conservation schemes must be followed in the area. Various ponds and reservoirs must be constructed in the area for the purpose of fisheries and providing boating facilities in the region.

WATERSHED MANAGEMENT

The problem of eco-management planning can be better understood with reference to watershed areas, and obviously the need for watershed management is vital for an efficient management of the eco-system in the region. It will include both the perennial as well as non-perennial segments of the river channels and land-locked water bodies (i.e. reservoirs, canals, tanks and ponds either natural or artificial in nature) of fluvially dominated terrain. The following suggestions may be postulated for the watershed management of the area:

(a) The major river basins like Denwa, Sonbhadra, Karighora, Bori and Nagdwari need the large scale management in their catchment areas. The tributaries of the major channels have their uncertain paths and the catchment areas found zigzag and strech in a haphazard way. Therefore, the identification of interfluves and water divides must be considered at first priority.

(b) The meandering flow path of rivers must be converted as a straight flaves otherwise, it may produce oxbow lakes. This type of conditions have been marked in the flow path of river Denwa, Nagdwari, Sonbhadra and Bori. It is also marked that the meandring flow path is totally governed by hard rocks making gorge positions and the flow paths show break in slope due to lithological controls associated with rapids and falls. Most of the channels of the area have been covered with boulders and thus the clearance of the channels are necessary for easy draining of water. If the few parts of the river Denwa have been thought to be mechanised in sense of the removal of the boulders, it would be better chance for developing the navigation facilities.

(c) The encroachment by the construction of pakka ghats and temples along the river Jambudeep near Jatashanker hills have consumed the width of the channel which in turn causes break in flowing the water. The unplanned constructions projected towards the valley must be prohibited.

(d) The quantity of surface water discharge in major channels must be made sufficient by removing the waste materials in the flow path.

(e) The chemical and bacteriological examination of river water reveals the fact that the water is hard and unsuitable for drinking purposes. The water-borne diseases and the impact of epidemics are the common factors caused by the polluted river water. Therefore, the medical treatment of water and cleaning up the surrounding areas are too much essential. It is found that the nala before the Military campus and near the Genesh hills which transport the sewerage of the town are too much dirty. The above sites may be maintained by the proper supply of water. The nala surrounding the Military campus is associated with pakka ghats. If it is maintained by providing water supply then it may be used as a boating club. It is the best place located in the heart of the town for the recreation of the tourists.

(f) The major waterfalls and rapids of the region are located in the upper part of the plateau rim. Waterfalls and rapids which are most common features of the area may be utilsed for the supply of drinking water.

(g) The ponds and reservoirs which are natural, artificial, perennial and non-perennial make the site of land-locked water bodies and require the watershed management. The region is rich in natural ponds. Ponds, just like small lake conditions are very much common along the Denwa river and it is also a very common features near the existence of rapids and waterfalls. The Pachmarhi reservoir (Van Shri Vihar) which is famous for boating must be developed in size and shape and facilities must be provided there in view of tourist's interest.

(h) The development of navigational facilities are the major objectives of watershed management but the channels of the region are not suitable. In the lower section of Denwa where it divorces the plateau region, navigation may be developed improving the depth and width of the river.

GROUNDWATER MANAGEMENT

Groundwater management is a significant issue in eco-management planning because it provides the socio-economic facilities and life-support system of the region. The management plan incorporates the possibilities of ground-water utilisation for existing population as well as future generation. The region is thickly dotted with springs and seepages. These springs are found every where in hilly tract and provide very cold water. Springs generate pond conditions and it is also seen that the local people also conserve the water from such springs by constructing small rectangular tanks. Springs are the major natural exposures of groundwater and the vital source of drinking water supply to the meagre population of the plateau. It is suggested to construct large scale concrete tanks at the sites of the springs and water supply must be managed by pipe system in the water scarcity zones. The small tanks and kunds beside the springs must be enlarged.

The groundwater of the region creates some health problems. In this connection, the maintenance of water quality standard, appointment of chemist under 'Jal Nigam', collection of water samples and their chemical treatment, establishment of chemical laboratory in Pachmarhi town, appropriate use of bleaching powder in well water etc. may help to maintain the quality of water in the region. To avoid pollution and to check the water-borne diseases one should adopt some precautions in the construction of wells such as location of well site grouting and scaling of well causing, completion of tap of well, disinfection of well, sanitary protection of pumping facilities, scaling the abandoned wells and moreover providing the facilities of Primary Health Centre in the problem area.

In view of provision of safe and adequate water supply of treated water, to minimise the rate of water-borne diseases, to quench the thirst of scarcity areas in summer, to improve the general health of the people and to stimulate the social and economic development, it is necessary to plan the water supply schemes through pipe system. The connections should be made easily available for the tribal population.

The problems of short period of water supply, operation and maintenance problems, uncertainity of power supply, lack of public acceptance and participation due to high cost of connection bills and water tax and the harassment by the officials of the 'Jal Sansthan' require the proper remedial actions. The supply of water in tribal areas must be made available by low monthly charges.

Pipe water supply scheme is a costly affair. Therefore, it is suggested that the scarcity areas should be covered by installing handpumps. Handpumps can be installded in rocky areas by sophisticated drilling rigs. The area survey for assessing the potentials of groundwater must be in practice.

The planning strategy must be made in social and commercial context. In order to save time and energy in relation to water collection journey and to improve rural health under the first stage benefits, the immediate aim should be the improvement of the quality, quantity, availability and reliability of water (Feacham, 1975).

TECHNO-MANAGEMENT FOR ECO-DECORATION, TERRAIN PROSPERITY AND TOURISM DEVELOPMENT

Techno-management includes all types of management schemes which are related to science and technology and totally governed by human activities. Man has fashioned the natural beauty providing some artificial means. The development of roads, railways, transport and telecommunication facilities, housing, hotelling and some other recreational tools, electric supply, water supply, post and telegraph, police station and security system, markets, banking and insurance, parking, gardening, shoping, boating, bathing, travelling etc. may be included in techno-management schemes of the area. Pachmarhi is well decorated showing fine natural beauty by itself. If it is further think to ornament the attractive corners of Pachmarhi, it will be necessary to manage some excellent facilities for this hill centre in view of tourism development, because it is the place of sweet dreams for the tourists. Taking into consideration of prosperous natural properties and historico-spiritual contents, activities of promoting tourism industry in the region is very essential. According to survey conducted in the area, it is apparent that the lodging and boarding facilities are unsatisfactory. The available lodging facilities are of high rating. Therefore few Dharamshalas may be arranged in minimum rates. The pitch roads may be constructed upto every tourists sites. Road lighting and 'Sulabh Shauchalaya' are the most essential needs. Market facilities may be improved. A ropeway may be arranged between Chauragarh and Dhupgarh. The number of taxies and bushes may be increased for the easy transportation in the area. The places like Dhupgarh, Chauragarh, Mahadeo and various other picnic spots may be fashioned with altra-modern facilities. During the fair season, the extra arrangement of fooding, lodging and transportation are too much essential.

Being a famous tourist-cum-religious centre, Pachmarhi hills is facing a number of acute problems. Warning for new unauthorised constructions, improvement of road-width in Parikrama marg, new plantation along the roads, public information centres, guide map and site indicators on road sides, reconstructions of old bridges, tamples, dharamshalas, roads and ghats, making suitable the navigational sites, arrangement of advanced boarding and lodging facilities, making available the market and other infrastructural facilities etc. are the essential infrastructures which the local authority should take care of in a planned way. The proposal of 'Van Vihar' including parks and sanctuaries are acceptable. The railway and air transport facilities are the most necessary needs for the region.

CONCLUSION

On the basis of the overall discussion, it may be concluded that the Pachmarhi hills needs some scientific management for the lovers of the area. It is an 'Eden of bliss' and so the beauty of the place must be decorated by all means. Naturally, it has left himself all the glamours of the nature, but in sense of modernisation and change, it needs few alarming facilities of techno-scientific yields, to recover and fortify the present figures alloted therein.

REFERENCES

1. Eckholm, E. 1978: *Disappearing species; The Social Challenge,* World Watch Paper-22.
2. Feacham, R. 1975: The Rational allocation of water resources for the domestic needs of rural communities in developing countries, Proc. of the Second World Congress on Water Resources, Vo1.2, Health and Planning, New Delhi, p.554.
3. Prasad, G. 1985: Environmental crisis of soil erosion on Chitrakut upland, Proceeding of the International Conference on Modelling

and Simulation held at Gorakhpur, India, Dec. 1985.

4. Prasad, G. 1987: Mode of formation and spatial analysis of major soils group of Chitrakut upland, India, *Modelling, Simulation and Control,* AMSE Press, France, C, Vol.7, No.4, pp.11-24.

5. Prasad, G. 1991: Morphological Processes, landforms and their significance in Environmental Management of Chitrakut and its adjoining region, Ph.D. thesis (unpublished) submitted to Kanpur University.

6. Singh, L.R., Singh, S., Tiwari, R.C. and Srivastava, R.P. 1983: *Environmental Management* (edited) Allahabad Geographical Society, Geographical Department, Allahabad University, p.361.

7. Singh, S. 1991: *Environmental Geography,* Prayag Pustak Bhawan, Allahabad, 1st Edition, p.464.

8. Vink, A.P.A. 1983: *Landscape Ecology and Landuse,* Longman Group Limited, London and New York.